METABOLIC PROBES OF
CENTRAL NERVOUS SYSTEM
ACTIVITY IN
EXPERIMENTAL ANIMALS AND MAN

JONATHAN MAGNES
1912–1980

The Magnes Lecture Series was established at the Hebrew University—Hadassah Medical School in memory of Professor Jonathan Magnes by his family, students, colleagues and friends. Jonathan Magnes is considered by many to be the founding father of Israeli physiology and neurobiology.

Magnes was born in New York and at the age of 10 immigrated to Israel where his father, Judah Magnes, was elected the first President of the Hebrew University of Jerusalem. Following graduation from high school, Jonathan Magnes decided to prepare himself for agricultural life in Palestine. After two years in the Geva settlement he was sent to study animal science in Davis, California. Upon his return, he became a key figure in the establishment of the milk industry. He introduced the mechanical milking process and built the first machine himself. This led to a very prosperous industry based on high yield and high hygienic quality.

The interest Magnes had in animals during his agricultural years led him to undertake doctoral work with Alexander Geiger. They developed the isolated brain preparation in which, by a meticulous surgical procedure, they were able to cannulate and ligate most of the blood vessels leading to and coming from the brain. Hence they were able to perfuse the brain and determine its metabolic balance. They arrived at a very important conclusion which seems almost trivial today but was of immense importance at that time: the main substance for brain function is the sugar glucose.

This brain perfusion preparation was Magnes's main experimental approach and resulted in a number of very interesting observations during the next three decades of experimental physiology. He studied the changes in brain metabolism during chemically induced convulsion, electroshock, barbiturate narcosis, and induction of sleep. In spite of his numerous teaching and administrative duties, he always performed experiments with his own hands, up until a month before his death. Tuesday was a "sacred" day for Magnes, devoted to experimentation. On one occasion, when forced to cancel an experiment, he commented, "A week without an experiment is aging without grace."

Magnes was one of the key figures of the Hebrew University—Hadassah Medical School from its creation in 1949. At the beginning he was in charge of finding temporary dwellings for the newly-founded school. Afterwards he was chairman of the building committee which planned new buildings at the Ein-Karem campus. Magnes was considered to be one of the great deans of the Hebrew University—Hadassah Medical School. He introduced many important innovations in teaching, administration, and organization of research.

The Department of Physiology at the Hadassah Medical School was, in his own mind, his major achievement. He created in the department a very special atmosphere, consisting of intense activity, friendship, critical discussion and intellectual honesty. But the most important element, to my mind, was the almost absolute scientific freedom. Every one of the young members of the department was allowed to pursue his scientific imagination with no restrictions (within the limits of the economic infrastructure). A few days before his death he mentioned that it gave him great pride that graduates from his department are professors in all four medical schools in Israel.

I hope this lecture series will commemorate this great architect of Israeli biomedical science in the spirit in which he lived and believed.

RAMI RAHAMIMOFF
Jerusalem

MAGNES LECTURE SERIES

METABOLIC PROBES OF CENTRAL NERVOUS SYSTEM ACTIVITY IN EXPERIMENTAL ANIMALS AND MAN

MAGNES LECTURE SERIES
VOLUME I

LOUIS SOKOLOFF

Laboratory of Cerebral Metabolism
National Institute of Mental Health

SINAUER ASSOCIATES INC. □ PUBLISHERS
SUNDERLAND, MA 01375

The author expresses his appreciation to Mr. J.D. Brown for his skilled photographic reproduction, and to Mrs. Ruth Bower for her excellent editorial and bibliographic assistance.

METABOLIC PROBES OF CENTRAL NERVOUS SYSTEM ACTIVITY IN EXPERIMENTAL ANIMALS AND MAN

Magnes Lecture Series

Volume I

Library of Congress Cataloging in Publication Data

Sokoloff, Louis, 1921–
    Metabolic probes of central nervous system activity in experimental animals and man.

    Includes bibliographies and index.
    1. Brain chemistry. 2. Metabolism. 3. Central
nervous system. I. Title. [DNLM: 1. Brain—metabolism.
WL 300 S683m]
QP376.S66   1984   599'.0188   84-5428
ISBN 0-87893-767-6 (pbk.)

Sinauer Associates Inc. ◻ Publishers

Sunderland, Massachusetts 01375

Printed in U.S.A.

5 4 3 2 1

CONTENTS

METABOLIC PROBES OF
CENTRAL NERVOUS SYSTEM
ACTIVITY IN
EXPERIMENTAL ANIMALS AND MAN

INTRODUCTION

The brain is a complex, heterogeneous organ composed of many anatomical and functional components. These components have markedly different levels of functional activity that vary independently with time and function. Other tissues are generally far more homogeneous, with most of their cells functioning similarly and synchronously in response to a common stimulus or regulatory influence. The central nervous system, however, consists of innumerable subunits, each integrated into its own set of functional pathways and networks and subserving only one or a few of the many activities in which the nervous system participates. Understanding how the nervous system functions requires knowledge, not only of the mechanisms of excitation and inhibition, but even more so of their precise localization in the nervous system and of the relationships of neural subunits to specific functions.

Historically, studies of the central nervous system have

1

concentrated heavily on localization of function and mapping of pathways related to specific functions. These have been carried out neuroanatomically and histologically with staining and degeneration techniques, behaviorally with ablation and stimulation techniques, electrophysiologically with electrical recording and evoked electrical responses, and histochemically with a variety of techniques, including fluorescence and immunofluorescence methods and autoradiography of orthograde and retrograde axoplasmic flow. Many of these conventional methods suffer from a sampling problem. They generally permit examination of only one potential pathway at a time, and only positive results are interpretable. Furthermore, the demonstration of a pathway reveals only a potential for function; it does not reveal its significance in normal function.

Tissues that do physical and/or chemical work, such as heart, kidney, and skeletal muscle, exhibit a close relationship between energy metabolism and the level of functional activity. The existence of a similar relationship in the tissues of the central nervous system has been more difficult to prove, partly because of uncertainty about the nature of the work associated with nervous functional activity, but mainly because of the difficulty in assessing the levels of functional and metabolic activities in the same functional component of the brain at the same time. Much of our present knowledge of cerebral energy metabolism in vivo has been obtained by means of the nitrous oxide technique of Kety and Schmidt (1948a) and its modifications (Scheinberg and Stead, 1949; Lassen and Munck, 1955; Eklöf et al., 1973; Gjedde et al., 1975), which measure the average rates of energy metabolism in the brain as a whole. These methods have demonstrated changes in cerebral metabolic rate in association with gross or diffuse alterations of cerebral function and/or structure, as, for example, those that occur during postnatal development, aging, senility, anesthesia, disorders of consciousness, and convulsive states (Kety, 1950, 1957; Lassen, 1959; Sokoloff, 1960, 1976). They have not detected changes in cerebral metabolic rate in a number of conditions with, perhaps, more subtle alterations in cerebral functional activity, for example, deep slow-wave sleep, performance of mental arithmetic, sedation and tranquilization, schizophrenia, and LSD-induced psychosis (Kety, 1950; Lassen, 1959; Sokoloff, 1969). It is possible that there are no changes in cerebral energy metabolism in these conditions. The apparent lack of change could also be explained either by a redistribution of local levels of functional and metabolic activity without significant change in the average activity of the brain as a whole or by the restriction of altered metabolic activity to regions too small to be detected in measurements of the brain as a whole. What has clearly been needed is a method that measures the rates of energy metabolism in specific, discrete regions of the brain in normal and in altered states of functional activity.

In pursuit of this goal Kety and his associates (Landau et al., 1955; Freygang and Sokoloff, 1958; Kety, 1960; Reivich et al., 1969) developed a quantitative autoradiographic technique to measure the local tissue concentrations of chemically inert, diffusible, radioactive tracers. They used this technique to determine the rates of blood flow simultaneously in all the structural components visible and identifiable in autoradiographs of serial sections of the brain. The application of this quantitative autoradiographic technique to the determination of local cerebral metabolic rate has proved to be more difficult because of the inherently greater complexity of the problem and the unsuitability of the labeled species of the normal substrates of cerebral energy metabolism: oxygen and glucose. The radioisotopes of oxygen have too short a physical half-life. Both oxygen and glucose are too rapidly converted to carbon dioxide, and CO_2 is too rapidly cleared from the cerebral tissues. Sacks (1957), for example, has found in humans significant losses of $^{14}CO_2$ from the brain within 2 minutes after the onset of an intravenous infusion of [^{14}C]glucose, labeled either uniformly or in the C-1, C-2, or C-6 positions. These limitations of [^{14}C]glucose have been avoided by the use of 2-deoxy-D-[^{14}C]glucose, a labeled analogue of glucose with special properties that make it particularly appropriate for this application (Sokoloff et al., 1977). It is metabolized through part of the pathway of glucose metabolism at a definable rate relative to that of glucose. Unlike glucose, however, its product—[^{14}C]deoxyglucose-6-phosphate—is essentially trapped in the tissues, allowing the application of the quantitative autoradiographic technique. The use of radioactive 2-deoxyglucose to trace glucose utilization and the autoradiographic technique to achieve regional localization has recently led to the development of a method that measures the rates of glucose utilization simultaneously in all components of the central nervous system in the normal, conscious state and during experimental physiological, pharmacological, and pathological conditions (Sokoloff et al., 1977). Because the procedure is so designed that the concentrations of radioactivity in the tissues during autoradiography are more or less proportional to the rates of glucose utilization, the autoradiographs provide pictorial representations of the relative rates of glucose utilization in all the cerebral structures visualized. Numerous studies with this method have established that there is a close relationship between functional activity and energy metabolism in the central nervous system (Sokoloff, 1977; Plum et al., 1976), and the method has become a potent new tool for mapping functional neural pathways on the basis of evoked metabolic responses.

THEORY

The method is derived from a model based on the biochemical properties of 2-deoxyglucose (Figure 1) (Sokoloff et al., 1977). 2-Deoxyglucose is transported bidirectionally between blood and brain by the same carrier that transports glucose across the blood–brain barrier (Bidder, 1968; Bachelard, 1971; Oldendorf, 1971). In the cerebral tissues it is phosphorylated by hexokinase to 2-deoxyglucose-6-phosphate (Sols and Crane, 1954). Deoxyglucose and glucose are, therefore, competitive substrates for both blood–brain transport and hexokinase-catalyzed phosphorylation. Unlike glucose-6-phosphate, however, which is further metabolized (eventually to CO_2 and water and to a lesser degree via the hexose monophosphate shunt), deoxyglucose-6-phosphate cannot be converted to fructose-6-phosphate and is not a substrate for glucose-6-phosphate dehydrogenase (Sols and Crane, 1954). There is very little glucose-6-phosphatase activity in brain (Hers, 1957) and even less deoxyglucose-6-phosphatase activity there (Sokoloff et al., 1977). Deoxyglucose-6-phosphate can be converted into deoxyglucose-1-phosphate and then into UDP-deoxyglucose and eventually into glycogen, glycolipids, and glycoproteins; but these reactions are slow, and in mammalian tissues only a very small fraction of the deoxyglucose-6-phosphate formed proceeds to these products (Nelson and Sokoloff, 1983). In any case, these represent products of deoxyglucose-6-phosphate and, therefore, relatively stable products of deoxyglucose phosphorylation. Deoxyglucose-6-phosphate and its derivatives, once formed, are, therefore, essentially trapped in the cerebral tissues, at least long enough for the duration of the measurement.

If the interval of time is kept short enough (for example,

PLASMA		BRAIN TISSUE	
		Precursor pool	Metabolic products

Glycolipids

$[^{14}C]$Deoxyglycogen

Glycoproteins

$[^{14}C]$UDPDG

$[^{14}C]$Deoxyglucose-1-phosphate

$[^{14}C]$Deoxyglucose (C_P^*) $\underset{k_2^*}{\overset{k_1^*}{\rightleftarrows}}$ $[^{14}C]$Deoxyglucose (C_E^*) $\xrightarrow{k_3^*}$ $[^{14}C]$Deoxyglucose-6-phosphate (C_M^*)

Total tissue ^{14}C concentration $= C_I^* = C_E^* + C_M^*$

BLOOD–BRAIN BARRIER

Glucose (C_P) $\underset{k_2}{\overset{k_1}{\rightleftarrows}}$ Glucose (C_E) $\xrightarrow{k_3}$ Glucose-6-phosphate (C_M)

$CO_2 + H_2O$

FIGURE 1

Diagrammatic representation of the theoretical model. C_i^* represents the total ^{14}C concentration in a single homogeneous tissue of the brain. C_P^* and C_P represent the concentrations of $[^{14}C]$deoxyglucose and glucose in the arterial plasma, respectively; C_E^* and C_E represent their respective concentrations in the tissue pools that serve as substrates for hexokinase. C_M^* represents the concentration of $[^{14}C]$deoxyglucose-6-phosphate in the tissue. The constants k_1^*, k_2^*, and k_3^* represent the rate constants for carrier-mediated transport of $[^{14}C]$deoxyglucose from plasma to tissue, for carrier-mediated transport back from tissue to plasma, and for phosphorylation by hexokinase, respectively. The constants k_1, k_2, and k_3 are the equivalent rate constants for glucose. $[^{14}C]$Deoxyglucose and glucose share and compete for the carrier that transports both between plasma and tissue and for hexokinase, which phosphorylates them to their respective hexose-6-phosphates. The dashed arrow represents the possibility of glucose-6-phosphate hydrolysis by glucose-6-phosphatase activity, if any. (From Sokoloff et al., 1977.)

less than 1 hour) to allow the assumption of negligible loss
of [^{14}C]DG-6-P from the tissues, then the quantity of [^{14}C]DG-
6-P accumulated in any cerebral tissue at any given time
following the introduction of [^{14}C] DG into the circulation is
equal to the integral of the rate of [^{14}C] DG phosphorylation
by hexokinase in that tissue during that interval of time. This
integral is in turn related to the amount of glucose that has
been phosphorylated over the same interval, depending on
the time courses of the relative concentrations of [^{14}C]DG and
glucose in the precursor pools and on the Michaelis-Menten
kinetic constants for hexokinase with respect to both [^{14}C]DG
and glucose. With cerebral glucose consumption in a steady
state, the amount of glucose phosphorylated during the interval
of time equals the steady-state flux of glucose through the
hexokinase-catalyzed step times the duration of the interval;
and the net rate of flux of glucose through this step equals
the rate of glucose utilization.

These relationships can be mathematically defined and
an operational equation derived if the following assumptions
are made: (1) a steady state for glucose (i.e., constant plasma
glucose concentration and constant rate of glucose consumption) throughout the period of the procedure; (2) homogeneous
tissue compartment within which the concentrations of
[^{14}C]DG and glucose are uniform and exchange directly with
the plasma; and (3) tracer concentrations of [^{14}C]DG (i.e.,
molecular concentrations of free [^{14}C]DG essentially equal to
zero). The operational equation that defines R_i, the rate of
glucose consumption per unit mass of tissue i in terms of
measurable variables, is presented in Figure 2.

The rate constants are determined in a separate group of
animals by a nonlinear, iterative process that provides the
least squares best-fit of an equation that defines the time
course of total tissue ^{14}C concentration in terms of the time,
the history of the plasma concentration, and the rate constants
to the experimentally determined time courses of tissue and

GENERAL EQUATION
FOR MEASUREMENT OF REACTION RATES WITH TRACERS:

$$\text{Rate of reaction} = \frac{\text{Labeled product formed in interval of time, 0 to } T}{\left[\begin{array}{c}\text{Isotope effect}\\\text{correction factor}\end{array}\right]\left[\begin{array}{c}\text{Integrated specific activity}\\\text{of precursor}\end{array}\right]}$$

OPERATIONAL EQUATION OF [^{14}C] DEOXYGLUCOSE METHOD:

Labeled product formed in interval of time, 0 to T

$$\overbrace{\qquad\qquad\qquad\qquad\qquad\qquad\qquad\qquad\qquad}$$

$$\underbrace{\begin{array}{c}\text{Total }^{14}\text{C in tissue}\\\text{at time } T\end{array}}\qquad\underbrace{\begin{array}{c}^{14}\text{C in precursor remaining in tissue}\\\text{at time } T\end{array}}$$

$$R_i = \frac{C_i^*(T) \quad - \quad k_1^*\, e^{-(k_2^* + k_3^*)T}\int_0^T C_P^*\, e^{(k_2^* + k_3^*)t}dt}{\left[\dfrac{\lambda \cdot V_m^* \cdot K_m}{\Phi \cdot V_m \cdot K_m^*}\right]\left[\displaystyle\int_0^T\left(\frac{C_P^*}{C_P}\right)dt \; - \; e^{-(k_2^* + k_3^*)T}\int_0^T\left(\frac{C_P^*}{C_P}\right)e^{(k_2^* + k_3^*)t}dt\right]}$$

| "Isotope effect" correction factor | Integrated plasma specific activity | Correction for lag in tissue equilibration with plasma |

Integrated precursor specific activity in tissue

FIGURE 2

Operational equation of radioactive deoxyglucose method and its functional anatomy. T represents the time at the termination of the experimental period; λ equals the ratio of the distribution space of deoxyglucose in the tissue to that of glucose; Φ equals the fraction of glucose that, once phosphorylated, continues down the glycolytic pathway; and K$_m^*$ and V$_m^*$ and K$_m$ and V$_m$ represent the familiar Michaelis-Menten kinetic constants of hexokinase for deoxyglucose and glucose, respectively. The other symbols are the same as those defined in Figure 1. (From Sokoloff, 1978a.)

plasma concentrations of ^{14}C (Sokoloff et al., 1977). The rate constants have thus far been completely determined only in normal, conscious albino rats (Table 1). Partial analyses indicate that the values are quite similar in the conscious monkey (Kennedy et al., 1978), dog (Duffy et al., 1982), and cat (M. Miyaoka, J. Magnes, C. Kennedy, and L. Sokoloff, unpublished data).

The λ, Φ, and enzyme kinetic constants are grouped together to constitute a single, lumped constant (Figure 2). It can be shown mathematically that this lumped constant is equal to the asymptotic value of the product of the ratio of the cerebral extraction ratios of $[^{14}C]DG$ and glucose and the ratio of the arterial blood to plasma specific activities when the arterial plasma $[^{14}C]DG$ concentration is maintained constant (Sokoloff et al., 1977). The lumped constant is also determined in a separate group of animals from arterial and cerebral venous blood samples drawn during a programmed intravenous infusion that produces and maintains a constant arterial plasma $[^{14}C]DG$ concentration (Sokoloff et al., 1977). An example of such a determination in a conscious monkey is illustrated in Figure 3. Thus far the lumped constant has been determined only in albino rat, monkey, cat, dog, and sheep (Table 2). The lumped constant appears to be characteristic of the species and does not appear to change significantly in a wide range of physiological conditions (Table 2) (Sokoloff et al., 1977). It has been found to change in pathophysiological conditions, such as severe hypoglycemia (Suda et al., 1981) and hyperglycemia (Schuier et al., 1981).

Despite its complex appearance, the operational equation is really nothing more than a general statement of the standard relationship by which rates of enzyme-catalyzed reactions are determined from measurements made with radioactive tracers (Figure 2). The numerator of the equation represents the amount of radioactive product formed in a given interval of time; it is equal to C_i^* (the combined concentrations of $[^{14}C]DG$

TABLE 1

VALUES OF RATE CONSTANTS
IN THE NORMAL CONSCIOUS ALBINO RAT

Structure	Rate constants ± standard error of estimates (min^{-1})		
	k_1^*	k_2^*	k_3^*
GRAY MATTER			
Visual cortex	0.189 ± 0.408	0.279 ± 0.176	0.063 ± 0.040
Auditory cortex	0.226 ± 0.068	0.241 ± 0.198	0.067 ± 0.057
Parietal cortex	0.194 ± 0.051	0.257 ± 0.175	0.062 ± 0.045
Sensory-motor cortex	0.193 ± 0.037	0.208 ± 0.112	0.049 ± 0.035
Thalamus	0.188 ± 0.045	0.218 ± 0.144	0.053 ± 0.043
Medial geniculate body	0.219 ± 0.055	0.259 ± 0.164	0.055 ± 0.040
Lateral geniculate body	0.172 ± 0.038	0.220 ± 0.134	0.055 ± 0.040
Hypothalamus	0.158 ± 0.032	0.226 ± 0.119	0.043 ± 0.032
Hippocampus	0.169 ± 0.043	0.260 ± 0.166	0.056 ± 0.040
Amygdala	0.149 ± 0.028	0.235 ± 0.109	0.032 ± 0.026
Caudate-putamen	0.176 ± 0.041	0.200 ± 0.140	0.061 ± 0.050
Superior colliculus	0.198 ± 0.054	0.240 ± 0.166	0.046 ± 0.042
Pontine gray matter	0.170 ± 0.040	0.246 ± 0.142	0.037 ± 0.033
Cerebellar cortex	0.225 ± 0.066	0.392 ± 0.229	0.059 ± 0.031
Cerebellar nucleus	0.207 ± 0.042	0.194 ± 0.111	0.038 ± 0.035
Mean	0.189	0.245	0.052
± SEM	± 0.012	± 0.040	± 0.010
WHITE MATTER			
Corpus callosum	0.085 ± 0.015	0.135 ± 0.075	0.019 ± 0.033
Genu of corpus callosum	0.076 ± 0.013	0.131 ± 0.075	0.019 ± 0.034
Internal capsule	0.077 ± 0.015	0.134 ± 0.085	0.023 ± 0.039
Mean	0.079	0.133	0.020
± SEM	± 0.008	± 0.046	± 0.020

Source: Sokoloff et al., 1977.

Structure	Distribution volume (mL/g) $k_1^*/(k_2^*+k_3^*)$	Half-life of precursor pool (min) $Log_e 2/(k_2^*+k_3^*)$
GRAY MATTER		
Visual cortex	0.553	2.03
Auditory cortex	0.734	2.25
Parietal cortex	0.608	2.17
Sensory-motor cortex	0.751	2.70
Thalamus	0.694	2.56
Medial geniculate body	0.697	2.21
Lateral geniculate body	0.625	2.52
Hypothalamus	0.587	2.58
Hippocampus	0.535	2.19
Amygdala	0.558	2.60
Caudate-putamen	0.674	2.66
Superior colliculus	0.692	2.42
Pontine gray matter	0.601	2.45
Cerebellar cortex	0.499	1.54
Cerebellar nucleus	0.892	2.99
Mean	0.647	2.39
± SEM	± 0.073	± 0.40
WHITE MATTER		
Corpus callosum	0.552	4.50
Genu of corpus callosum	0.507	4.62
Internal capsule	0.490	4.41
Mean	0.516	4.51
± SEM	± 0.171	± 0.90

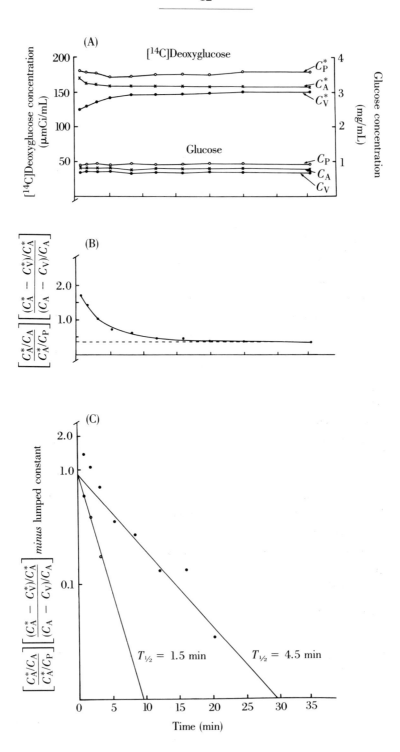

13

FIGURE 3

*Data obtained and their use in determination of the lumped
constant and the combination of rate constants (k_2^* + k_3^*) in
a representative experiment in a monkey. (A) Time courses of
arterial blood and plasma concentrations of [^{14}C] deoxyglucose
and glucose and cerebral venous blood concentrations of
[^{14}C] deoxyglucose and glucose during programmed intravenous
infusion of [^{14}C] deoxyglucose. (B) Arithmetic plot of the function derived from the variables in A and combined as indicated
in the formula on the ordinate against time. This function
declines exponentially, with a rate constant equal to (k_2^* +
k_3^*), until it reaches an asymptotic value equal to the lumped
constant (0.35 in this experiment—dashed line). (C) Semi-
logarithmic plot of the curve in B less the lumped constant,
i.e., its asymptotic value. Solid circles represent actual values.
This curve is analyzed into two components by a standard curve-
peeling technique to yield the two straight lines representing
the separate components. Open circles are points for the fast
component, obtained by subtracting the values for the slow
component from the solid circles. The rate constants for these
two components represent the values of (k_2^* + k_3^*) for two compartments; the fast and slow compartments are assumed to represent gray and white matter, respectively. In this experiment
the values for (k_2^* + k_3^*) were found to equal 0.462 (half-time
= 1.5 min) and 0.154 (half-time = 4.5 min) in gray and
white matter, respectively. (From Kennedy et al., 1978.)*

and [^{14}C]DG-6-P in the tissue at time T, measured by the
quantitative autoradiographic technique) less a term that represents the free unmetabolized [^{14}C]DG still remaining in the
tissue. The denominator represents the integrated specific
activity of the precursor pool times a factor (the lumped constant) that is equivalent to a correction factor for an isotope
effect. The term with the exponential factor in the denominator

TABLE 2

VALUES OF THE LUMPED CONSTANT
IN SEVERAL SPECIES

Animal	Number of animals	Mean ± SD	SEM
Albino Rat			
Conscious	15	0.464 ± 0.099^a	± 0.026
Anesthetized	9	0.512 ± 0.118^a	± 0.039
Conscious (5% CO_2)	2	0.463 ± 0.122^a	± 0.086
Combined	26	0.481 ± 0.119	± 0.023
Rhesus monkey			
Conscious	7	0.344 ± 0.095	± 0.036
Cat			
Anesthetized	6	0.411 ± 0.013	± 0.005
Dog (beagle puppy)			
Conscious	7	0.558 ± 0.082	± 0.031
Sheep			
Fetus	5	0.416 ± 0.031	± 0.014
Newborn	4	0.382 ± 0.024	± 0.012
Mean	9	0.400 ± 0.033	± 0.011

Sources: Rat (Sokoloff et al., 1977); monkey (Kennedy et al., 1978); cat (M. Miyaoka, J. Magnes, C. Kennedy, M. Shinohara, and L. Sokoloff, unpublished data); dog (Duffy et al., 1982); sheep (Abrams et al., 1984).

[a] No statistically significant difference between normal conscious and anesthetized rats ($0.3 < p < 0.4$) and conscious rats breathing 5% CO_2 ($p > 0.9$).

takes into account the lag in the equilibration of the tissue precursor pool with the plasma.

EXPERIMENTAL PROCEDURE FOR MEASUREMENT OF LOCAL CEREBRAL GLUCOSE UTILIZATION

Theoretical considerations in the design of the procedure

The operational equation of the method specifies the variables to be measured in order to determine R_i, the local rate of glucose consumption in the brain. The following variables are measured in each experiment: (1) the entire history of the arterial plasma [^{14}C]deoxyglucose concentration, C_P^*, from zero time to the time of killing T; (2) the steady-state arterial plasma glucose level C_P over the same interval; and (3) the local concentration of ^{14}C in the tissue at the time of killing, $C_i^*(T)$. The rate constants, k_1^*, k_2^*, and k_3^*, and the lumped constant, $\lambda V_m^* K_m/\Phi V_m K_m^*$, are not measured in each experiment; the values for these constants that are used are those determined separately in other groups of animals as described earlier and presented in Tables 1 and 2.

The operational equation is generally applicable with all types of arterial plasma [^{14}C]DG concentration curves. Its configuration, however, suggests that a declining curve approaching zero by the time of killing is the choice to minimize certain potential errors. The quantitative autoradiographic technique measures only total ^{14}C concentration in the tissue and does not distinguish between [^{14}C]DG-6-P and [^{14}C]DG. It is, however, [^{14}C]DG-6-P concentration that must be known to determine glucose consumption. [^{14}C]DG-6-P concentration is calculated in the numerator of the operational equation, which equals the total tissue ^{14}C content, $C_i^*(T)$, minus the [^{14}C]DG concentration present in the tissue, estimated by the term containing the exponential factor and rate constants. In the denominator of the operational equation, there is also a term containing an exponential factor and rate constants. Both these terms have the useful property of approaching zero with increasing time if C_P^* is also allowed to approach zero. The rate constants, k_1^*, k_2^*, and k_3^*, are not measured in the same animals in which local glucose consumption is being measured. It is conceivable that the rate constants in Table 1 are not equally applicable in all physiological, pharmacological, and pathological states. One possible solution is to determine the rate constants for each condition to be studied. An alternative solution, and the one chosen, is to administer the [^{14}C]DG as a single intravenous pulse at zero time and to allow sufficient time for the clearance of [^{14}C]DG from the plasma and the terms containing the rate constants to fall to levels too low to influence the final result. To wait until these terms reach zero is impractical because of the long time required and the risk of effects of the small but finite rate of loss of [^{14}C]DG-6-P from the tissues. A reasonable time interval is 45 minutes; by this time the plasma level has fallen to very low levels, and, on the basis of the values of $(k_2^* + k_3^*)$ in Table 1, the exponential factors have declined through at least ten half-lives, at least under physiological conditions.

Experimental protocol

The animals are prepared for the experiment by the insertion of polyethylene catheters in an artery and a vein. Any convenient artery and vein can be used. In the rat the femoral or the tail arteries and veins have been found satisfactory. In the monkey and cat the femoral vessels are probably most convenient. The catheters are inserted under anesthesia, and anesthetic agents without long-lasting aftereffects should be

used. Light halothane anesthesia with or without supplemen-
tation with nitrous oxide have been found to be quite satisfactory. At least 2 hours are allowed for recovery from the
surgery and anesthesia before initiation of the experiment.

The design of the experimental procedure for the measurement of local cerebral glucose utilization was based on
the theoretical considerations discussed earlier. At zero
time a pulse of 125 μCi (no more than 2.5 μmoles) of
[^{14}C]deoxyglucose per kilogram of body weight is administered
to the animal via the venous catheter. Arterial sampling is
initiated with the onset of the pulse, and timed, 50 to 100-
μL samples of arterial blood are collected consecutively as
rapidly as possible during the early period so as not to miss
the peak of the arterial curve. Arterial sampling is continued
at less frequent intervals later in the experimental period but
at sufficient frequency to define fully the arterial curve. The
arterial blood samples are immediately centrifuged to separate
the plasma, which is stored on ice until assayed for [^{14}C]DG
concentrations by liquid scintillation counting and glucose
concentrations by standard enzymatic methods. At approximately 45 minutes the animal is decapitated, the brain is
removed and frozen in Freon XII or isopentane maintained
between $-50°$ and $-75°C$ with liquid nitrogen. When fully
frozen, the brain is stored at $-70°C$ until sectioned and au-
toradiographed. The experimental period may be limited to
30 minutes. This is theoretically permissible and may sometimes be necessary for reasons of experimental expediency,
but greater errors due to possible inaccuracies in the rate
constants may result.

Autoradiographic measurement of tissue ^{14}C concentration

The ^{14}C concentrations in localized regions of the brain are
measured by a modification of the quantitative autoradi-

ographic technique previously described (Reivich et al., 1969). The frozen brain is coated with chilled embedding medium (Lipshaw Manufacturing Co., Detroit, MI) and fixed to object-holders appropriate to the microtome to be used.

Brain sections, precisely 20 μm in thickness, are prepared in a cryostat maintained at $-21°$ to $-22°C$. The brain sections are picked up on glass cover slips, dried on a hot plate at 60°C for at least 5 minutes, and placed sequentially in an X-ray cassette. A set of [^{14}C]methylmethacrylate standards (Amersham Corp., Arlington Heights, IL), which include a blank and a series of progressively increasing ^{14}C concentrations, are also placed in the cassette. These standards must previously have been calibrated for their autoradiographic equivalence to the ^{14}C concentrations in brain sections (20 μm in thickness) prepared as described above. The method of calibration has been previously described (Reivich et al., 1969).

Autoradiographs are prepared from these sections directly in the X-ray cassette with Kodak single-coated, blue-sensitive Medical X-ray Film, Type SB-5 (Eastman Kodak Co., Rochester, NY). The exposure time is generally 5–6 days with the doses used as described above, and the exposed films are developed according to the instructions supplied with the film. The SB-5 X-ray film is rapid but coarse grained. For finer-grained autoradiographs and, therefore, better-defined images with higher resolution, it is possible to use mammographic films, such as DuPont LoDose or Kodak MR-1 films, or fine-grain panchromatic film, such as Kodak Plus-X, but the exposure times are two to three times longer. The autoradiographs provide a pictorial representation of the relative ^{14}C concentrations in the various cerebral structures and the plastic standards. A calibration curve of the relationship between optical density and tissue ^{14}C concentration for each film is obtained by densitometric measurements of the portions of the film representing the various standards. The local tissue concentrations are then determined from the calibration curve and the optical densities of the film in the regions representing the cerebral structures of interest. Local cerebral glucose utilization is calculated from the local tissue concentrations of ^{14}C and the plasma [^{14}C]DG and glucose concentrations according to the operational equation (Figure 2).

Theoretical and practical considerations

The design of the deoxyglucose method is based on an operational equation, derived by the mathematical analysis of a model of the biochemical behavior of [^{14}C]deoxyglucose and glucose in brain (Figures 1 and 2). Although the model and its mathematical analysis are as rigorous and comprehensive as reasonably possible, it must be recognized that models almost always represent idealized situations and cannot possibly take into account all the known, let alone unknown, properties of a complex biological system. Several years have now passed since the introduction of the deoxyglucose method, and numerous applications of it have been made. The results of this experience generally establish the validity and worth of the method, but there are some potential problems which might occur in special situations and which require further theoretical and practical consideration.

The main potential sources of error are the rate constants and the lumped constant. The problem with them is that they are not determined in the same animals and at the same time when local cerebral glucose utilization is being measured. They are measured in separate groups of comparable animals and then used subsequently in other animals in which glucose utilization is being measured. The part played by these constants in the method is defined by their role in the operational equation of the method (Figure 2).

RATE CONSTANTS   The rate constants k_1^*, k_2^*, and k_3^* for deoxyglucose have been fully determined for various cerebral tissues in the normal, conscious albino rat (Sokoloff et al.,

1977) (Table 1), but they appear to be similar in other species. All the rate constants vary considerably from tissue to tissue, but the variation among gray structures and among white structures is considerably less than the differences between the two types of tissues (Table 1). The rate constants k_2^* and k_3^* appear in the equation only as their sum, and (k_2^* + k_3^*) is equal to the rate constant for the turnover of the free [^{14}C]deoxyglucose pool in the tissue. The half-life of the free [^{14}C]deoxyglucose pool can then be calculated by dividing (k_2^* + k_3^*) into the natural logarithm of 2 and has been found to average 2.4 minutes in gray matter and 4.5 minutes in white matter in the normal, conscious rat (Table 1).

The rate constants vary not only from structure to structure but can be expected to vary with the condition. For example, k_1^* and k_2^* are influenced by both blood flow and transport of [^{14}C]deoxyglucose across the blood–brain barrier; and because of the competition for the transport carrier, the glucose concentrations in the plasma and tissue affect the transport of [^{14}C]deoxyglucose and, therefore, also k_1^* and k_2^*. The constant k_3^* is related to phosphorylation of [^{14}C]deoxyglucose and will certainly change when glucose utilization is altered. To minimize potential errors due to inaccuracies in the values of the rate constants used, it was decided to sacrifice time resolution for accuracy. If the [^{14}C]deoxyglucose is given as an intravenous pulse and sufficient time is allowed for the plasma to be cleared of the tracer, then the influence of the rate constants, and the functions that they represent, on the final result diminishes with increasing time until ultimately it becomes zero. This relationship is implicit in the structure of the operational equation (Figure 2); as C_P^* approaches zero, the terms containing the rate constants also approach zero with increasing time. The significance of this relationship is graphically illustrated in Figure 4. From typical arterial plasma [^{14}C]deoxyglucose and glucose concentration curves obtained in a normal, conscious rat, the portion of the denom-

OPERATIONAL EQUATION

$$R_i = \frac{C_i^*(T) - k_1^* \, e^{-(k_2^* + k_3^*)T} \int_0^T C_P^* \, e^{(k_2^* + k_2^*)t} dt}{\left[\dfrac{\lambda \cdot V_m^* \cdot K_m}{\Phi \cdot V_m^* \cdot K_m^*}\right]\left[\underline{\int_0^T \left(\dfrac{C_p^*}{C_p}\right) dt - e^{-(k_2^* + k_3^*)T} \int_0^T \left(\dfrac{C_p^*}{C_p}\right) e^{(k_2^* + k_3^*)t} dt}\right]}$$

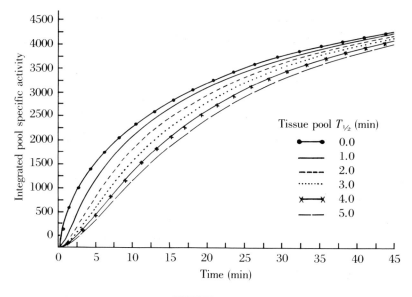

FIGURE 4

Influence of time and rate constants $(k_2^* + k_3^*)$ *on integrated precursor pool specific activity in a normal, conscious rat given an intravenous pulse of 50 μCi of $[^{14}C]$deoxyglucose at zero time. The time courses of the arterial plasma $[^{14}C]$deoxyglucose and glucose concentrations were measured following the pulse. The portion of the equation underlined (corresponding to integrated pool specific activity) was computed as a function of time with different values of $(k_2^* + k_3^*)$, as indicated by their equivalent half-lives, calculated according to $T\frac{1}{2} = 0.693/(k_2^* + k_3^*)$. (From Sokoloff, 1979.)*

inator of the operational equation underlined by the heavy bar was computed with a wide range of values for $(k_2^* + k_3^*)$ as a function of time. The values for $(k_2^* + k_3^*)$ are presented as their equivalent half-lives, which are calculated as described above. The values of $(k_2^* + k_3^*)$ vary from infinity (i.e., $T^{1/2} = 0$ minutes to 0.14 per minute (i.e., $T^{1/2} = 5$ minutes) and more than cover the range of values to be expected under physiological conditions. The portion of the equation underlined and computed represents the integral of the precursor pool specific activity in the tissue. The curves represent the time course of this function, one each for every value of $(k_2^* + k_3^*)$ examined. It can be seen that these curves are widely different at early times but converge with increasing time until at 45 minutes the differences over the entire range of $(k_2^* + k_3^*)$ equal only a small fraction of the value of the integral. These curves demonstrate that at short times enormous errors can occur if the values of the rate constants are not precisely known, but only negligible errors occur at 45 minutes, even over a wide range of rate constants of several fold. In fact, it was precisely for this reason that [14C]deoxyglucose rather than [14C]glucose was selected as the tracer for glucose metabolism. The relationships are similar for glucose. Because the products of [14C]glucose metabolism are so rapidly lost from the tissues, it is necessary to limit the experimental period to short intervals during which enormous errors can occur if the rate constants are not precisely known for each individual structure. [14C]Deoxyglucose permits the prolongation of the experimental period to intervals during which inaccuracies in rate constants have little effect on the final result.

It should be noted, however, that in pathological conditions the rate constants may fall far below the range examined in Figure 4. There is evidence, for example, that this occurs with hyperglycemia and ischemia (Hawkins et al., 1981). In such abnormal conditions it is necessary and feasible to redetermine the rate constants for the particular condition under study.

LUMPED CONSTANT   The lumped constant is composed of six separate constants. One of these, Φ, is a measure of the steady-state hydrolysis of glucose-6-phosphate to free glucose and phosphate. Because in normal brain tissue there is little such phosphohydrolase activity (Hers, 1957), Φ is normally approximately equal to unity. The other components are arranged in three ratios: λ, which is the ratio of distribution spaces in the tissue for deoxyglucose and glucose; V_m^*/V_m; and K_m/K_m^*. Although each individual constant may vary from structure to structure and condition to condition, it is likely that the ratios tend to remain the same under normal conditions. For reasons described in detail previously (Sokoloff et al., 1977), it is reasonable to believe that the lumped constant is the same throughout the brain and more or less characteristic of the species of animal under normal physiological conditions. Empirical experience thus far indicates that it is generally so, except in special pathophysiological states. The greatest experience has been accumulated in the albino rat. In this species the lumped constant for the brain as a whole has been determined under a variety of conditions (Sokoloff et al., 1977). In the normal, conscious rat, local cerebral glucose utilization, determined by the [^{14}C]deoxyglucose method with the single value of the lumped constant for the brain as a whole, correlates almost perfectly ($r = 0.96$) with local cerebral blood flow, measured by the [^{14}C]iodoantipyrine method, an entirely independent method (Sokoloff, 1978b). It is generally recognized that local blood flow is adjusted to local metabolic rate, but if the single value of the lumped constant did not apply to the individual structures studied, then errors in local glucose utilization would occur that might be expected to obscure the correlation. Also, the lumped constant has been directly determined in the albino rat in the normal conscious state, under barbiturate anesthesia, and during the inhalation of 5 percent CO_2; no significant differences were observed (Table 2). The lumped constant does vary with the species of animal. It has now also been determined in the rhesus monkey (Kennedy et al., 1978), cat (M. Miyaoka, J. Magnes, C. Kennedy, M. Shinohara, and L. Sokoloff, unpublished data), beagle puppy (Duffy et al., 1982), and sheep (Abrams, et al., 1984); and each species has a different value (Table 2). The values for local rates of glucose utilization determined with these lumped constants in these species are very close to what might be expected from measurement of energy metabolism in the brain as a whole by other methods (Table 3).

Although there is yet no experimental evidence of more than negligible changes in the lumped constant under physiological conditions, it certainly does change in pathophysiological states. In severe hypoglycemia there is a progressive and appreciable increase in the lumped constant with falling plasma glucose concentration (Suda et al., 1981); and in severe hyperglycemia there is a moderate decrease (Schuier et al., 1981). Tissue damage may also disrupt the normal cellular compartmentation, and there is no assurance that λ, the ratio of the distribution spaces for [^{14}C]deoxyglucose and glucose, is the same in damaged tissue as it is in normal tissue. In pathological states there may be release of lysosomal acid hydrolases that may hydrolyze glucose-6-phosphate and thus alter the value of Φ. It is necessary, therefore, to determine the lumped constant in each pathological or pathophysiological state.

GLUCOSE-6-PHOSPHATASE Although relatively low, the activity of glucose-6-phosphatase in brain is not zero (Sokoloff, 1982). This enzyme is capable of hydrolyzing the product, [^{14}C]deoxyglucose-6-phosphate, back to free [^{14}C]deoxyglucose and, thus, of causing loss of product. Fortunately, the product and enzyme are in separate cellular compartments. The

TABLE 3

RATES OF GLUCOSE UTILIZATION
IN REPRESENTATIVE STRUCTURES
OF THE BRAIN OF CONSCIOUS
ALBINO RAT, RHESUS MONKEY, AND CAT[a]

Structures	Rat (10)[a]	Cat (4)[c]	Monkey (7)[d]
Cerebral Cortex			
Striate	107 ± 4	95 ± 9	59 ± 2
Parietal	111 ± 5	69 ± 8	47 ± 4
Auditory	162 ± 5	118 ± 4	79 ± 4
Sensorimotor	120 ± 5	64 ± 3	44 ± 3
Thalamus			
Ventral posterolateral nucleus	91 ± 8	66 ± 5	43 ± 2
Pulvinar	96 ± 5	61 ± 5	41 ± 1
Medial geniculate body	131 ± 5	106 ± 14	65 ± 3
Lateral geniculate body	96 ± 5	109 ± 9	39 ± 1
Hypothalamus			
Preoptic area	57 ± 5	41 ± 4	26 ± 4
Medial mamillary nucleus	121 ± 5	85 ± 3	57 ± 3
Limbic System			
Amygdala	52 ± 2	34 ± 4	25 ± 2
Hippocampal gyrus	79 ± 3	57 ± 5	39 ± 2
Septal nucleus	64 ± 3	39 ± 3	26 ± 2
Basal Ganglia			
Caudate	110 ± 4	77 ± 2	52 ± 3
Globus pallidus	58 ± 2	36 ± 2	26 ± 2
Subthalamic nucleus	78 ± 5	61 ± 6	51 ± 2
Substantia nigra	58 ± 3	40 ± 5	29 ± 3
Red nucleus	68 ± 6	64 ± 7	48 ± 3
Brain Stem			
Superior colliculus	95 ± 5	82 ± 8	55 ± 4
Inferior colliculus	197 ± 10	171 ± 20	103 ± 6
Vestibular nucleus	128 ± 5	81 ± 7	66 ± 3

TABLE 3 (Continued)

Structures	Rat (10)[a]	Cat (4)[c]	Monkey (7)[d]
Pontine reticular nucleus	62 ± 3	46 ± 3	28 ± 1
Cerebellum			
Cortex	57 ± 2	73 ± 12	31 ± 2
Dentate nucleus	100 ± 4	68 ± 4	45 ± 2
White Matter			
Corpus callosum	40 ± 2	15 ± 2	11 ± 0
Internal capsule	33 ± 2	17 ± 2	13 ± 1

[a] Values are means ± standard errors obtained in the number of animals indicated in parentheses. Units of values for rates of glucose utilization are μmoles per 100 g per minute.

[b] From Sokoloff et al. (1977).

[c] From M. Miyaoka, J. Magnes, C. Kennedy, and L. Sokoloff (unpublished data).

[d] From Kennedy et al. (1978).

$[^{14}C]$DG-6-P is formed in the cytosol and must be transported into the cisterns of the endoplasmic reticulum, where the glucose-6-phosphatase resides, before the hydrolysis can occur. This compartmentalization provides a period of time before the effects of glucose-6-phosphatase become significant. When the experimental period is kept within 45 minutes, the influence of glucose-6-phosphatase is negligible and can be ignored (Sokoloff, 1982; Sokoloff et al., 1977). When, however, it is necessary to extend the experimental period to longer intervals, as, for example, in studies in humans with positron emission tomography (Phelps et al., 1979), it is necessary to account for the effects of glucose-6-phosphatase activity. This has been done by modifying slightly the original model to include k_4^* (the rate constant for $[^{14}C]$DG-6-P hydrolysis by glucose-6-phosphatase) and by deriving a modified operational equation that incorporates k_4^* (Huang et al., 1980; Sokoloff, 1982).

Computerized color-coded
image processing

The autoradiographs provide pictorial representations of only the relative concentrations of the isotope in the various tissues. Because of the use of a pulse followed by a long period before killing, the isotope is contained mainly in deoxyglucose-6-phosphate, which reflects the rate of glucose metabolism. The autoradiographs are, therefore, pictorial representations also of the relative, but not the actual, rates of glucose utilization in all the structures of the nervous system. Resolution of differences in relative rates, however, is limited by the ability of the human eye to recognize differences in shades of gray. Manual densitometric analysis permits the computation of actual rates of glucose utilization with a fair degree of resolution, but it generates enormous tables of data, which fail to convey the tremendous heterogeneity of local cerebral metabolic rates, even within anatomic structures, or the full information contained within the autoradiographs. Goochee et al. (1980) have developed a computerized image processing system to analyze and transform the autoradiographs into color-coded maps of the distribution of the actual rates of glucose utilization exactly where they are located throughout the central nervous system. The autoradiographs are scanned automatically by a computer-controlled scanning microdensitometer. The optical density of each spot in the autoradiograph (from 25 to 100 μm, as selected) is stored in a computer, converted to ^{14}C concentration on the basis of the optical densities of the calibrated ^{14}C plastic standards, and then converted to local rates of glucose utilization by solution of the operational equation of the method. Colors are assigned to narrow ranges of the rates of glucose utilization, and the autoradiographs are then displayed in a color TV monitor in color along with a calibrated color scale for identifying the rate of glucose utilization in each spot of the autoradiograph from its color. These color maps add a third dimension—the rate of glucose utilization on a color scale—to the spatial dimensions already present on the autoradiographs (see Figure 9).

RATES OF LOCAL, CEREBRAL GLUCOSE UTILIZATION IN THE NORMAL, CONSCIOUS STATE

Local rates of cerebral glucose utilization determined by the 2-[^{14}C]deoxyglucose method in normal, conscious, adult animals are presented in Table 3. The rates of local, cerebral glucose utilization in the normal, conscious rat vary widely throughout the brain. The values in white structures tend to group together and are always considerably below those of gray structures. The average value in gray matter is approximately three times that of white matter, but the individual values vary from approximately 50 to 200 μmoles glucose • 100 g^{-1} • min^{-1}. The highest values are in the structures involved in auditory functions, with the inferior colliculus clearly the most metabolically active structure in the brain.

The rates of local cerebral glucose utilization in the conscious monkey exhibit similar heterogeneity, but they are generally one-third to one-half the values in corresponding structures of the rat brain (Table 3). The differences in rates in rat and monkey brains are consistent with the different cellular packing densities in the brains of these two species.

Quantitative studies in normal, adult cats have also been carried out in collaboration with Dr. J. Magnes when he was a visiting scientist in our laboratory. The distribution of the local rates of glucose utilization are similar to those of the rat and monkey, but the values in the cat fall between those found in these two species (Table 3).

EFFECTS OF
GENERAL ANESTHESIA

General anesthesia produced by thiopental reduces the rates of glucose utilization in all structures of the rat brain (Table 4) (Sokoloff et al., 1977). The effects are not uniform, however. The greatest reductions occur in the gray structures, particularly those of the primary sensory pathways. The effects in white matter, though definitely present, are relatively small compared to those of gray matter. These results are in agreement with those of previous studies in which anesthesia has been found to decrease the cerebral metabolic rate of the brain as a whole (Kety, 1950; Lassen, 1959; Sokoloff, 1976).

TABLE 4

EFFECTS OF THIOPENTAL ANESTHESIA
ON LOCAL CEREBRAL GLUCOSE UTILIZATION
IN THE RAT[a]

Structure	Local cerebral glucose utilization ($\mu moles \cdot 100\ g^{-1} \cdot min^{-1}$)		
	Control (6)[b]	Anesthetized (8)[b]	Effect (%)
GRAY MATTER			
Visual cortex	111 ± 5	64 ± 3	− 42
Auditory cortex	157 ± 5	81 ± 3	− 48
Parietal cortex	107 ± 3	65 ± 2	− 39
Sensory-motor cortex	118 ± 3	67 ± 2	− 43
Lateral geniculate body	92 ± 2	53 ± 3	− 42
Medial geniculate body	126 ± 6	63 ± 3	− 50

TABLE 4 (Continued)

Structure	Local cerebral glucose utilization $(\mu moles \cdot 100\ g^{-1} \cdot min^{-1})$		
	Control (6)[b]	Anesthetized (8)[b]	Effect (%)
GRAY MATTER			
Thalamus: lateral nucleus	108 ± 3	58 ± 2	− 46
Thalamus: ventral nucleus	98 ± 3	55 ± 1	− 44
Hypothalamus	63 ± 3	43 ± 2	− 32
Caudate-putamen	111 ± 4	72 ± 3	− 35
Hippocampus: Ammon's horn	79 ± 1	56 ± 1	− 29
Amygdala	56 ± 4	41 ± 2	− 27
Cochlear nucleus	124 ± 7	79 ± 5	− 36
Lateral lemniscus	114 ± 7	75 ± 4	− 34
Inferior colliculus	198 ± 7	131 ± 8	− 34
Superior olivary nucleus	141 ± 5	104 ± 7	− 26
Superior colliculus	99 ± 3	59 ± 3	− 40
Vestibular nucleus	133 ± 4	81 ± 4	− 39
Pontine gray matter	69 ± 3	46 ± 3	− 33
Cerebellar cortex	66 ± 2	44 ± 2	− 33
Cerebellar nucleus	106 ± 4	75 ± 4	− 29
WHITE MATTER			
Corpus callosum	42 ± 2	30 ± 2	− 29
Genu of corpus callosum	35 ± 5	30 ± 2	− 14
Internal capsule	35 ± 2	29 ± 2	− 17
Cerebellar white matter	38 ± 2	29 ± 2	− 24

Source: From Sokoloff et al. (1977).

[a] Determined at 30 minutes following pulse of [^{14}C]deoxyglucose.

[b] The values are the means ± standard errors obtained in the number of animals indicated in parentheses. All the differences are statistically significant at the p 0.05 level.

RELATION BETWEEN
LOCAL FUNCTIONAL ACTIVITY
AND ENERGY METABOLISM

The results of a variety of applications of the method demonstrate a clear relationship between local, cerebral functional activity and glucose consumption. The most striking demonstrations of the close coupling between function and energy metabolism are seen with experimentally induced local alterations in functional activity that are restricted to a few specific areas in the brain. The effects on local glucose consumption are then so pronounced that they are not only observed in the quantitative results but can be visualized directly on the autoradiographs, which are really pictorial representations of the relative rates of glucose utilization in the various structural components of the brain.

Effects of increased functional activity

EFFECTS OF SCIATIC NERVE STIMULATION   Electrical stimulation of one sciatic nerve in the rat under barbiturate anesthesia causes pronounced increases in glucose consumption (i.e., increased optical density in the autoradiographs) in the ipsilateral dorsal horn of the lumbar spinal cord (Kennedy et al., 1975).

EFFECTS OF EXPERIMENTAL FOCAL SEIZURES   The local injection of penicillin into the hand–face area of the motor cortex of the rhesus monkey has been shown to induce electrical discharges in the adjacent cortex and to result in recurrent focal seizures involving the face, arm, and hand on the contralateral side (Caveness, 1969). Such seizure activity causes selective increases in glucose consumption in areas of motor cortex adjacent to the penicillin locus and in small discrete regions of the putamen, globus pallidus, caudate nucleus, thalamus, and substantia nigra of the same side (Figure 5) (Kennedy et al., 1975). Similar studies in the rat have led to comparable results and provided evidence on the basis of an evoked metabolic response of a "mirror" focus in the motor cortex contralateral to the penicillin-induced epileptogenic focus (Collins et al., 1976).

Effects of decreased functional activity

Decrements in functional activity result in reduced rates of glucose utilization. These effects are particularly striking in the auditory and visual systems of the rat and the visual system of the monkey.

EFFECTS OF AUDITORY OCCLUSION   In the albino rat some of the highest rates of local cerebral glucose utilization are found in components of the auditory system, that is, auditory cortex, medial geniculate nucleus, inferior colliculus, lateral lemniscus, superior olive, and cochlear nucleus (Table 3). Bilateral auditory deprivation by occlusion of both external auditory canals with wax markedly depresses the metabolic activity in all of these areas (Sokoloff, 1977). The reductions are symmetrical bilaterally and range from 35 to 60 percent. Unilateral auditory deprivation also depresses the glucose consumption of these structures, but to a lesser degree; and some of the structures are asymmetrically affected. For example, the metabolic activity of the ipsilateral cochlear nucleus equals 75 percent of the activity of the contralateral nucleus. The lateral lemniscus, superior olive, and medial geniculate ganglion are slightly lower on the contralateral side, whereas the contralateral inferior colliculus is markedly lower in metabolic activity than the ipsilateral structure. These results demonstrate that there is some degree of lateralization and crossing of auditory pathways in the rat.

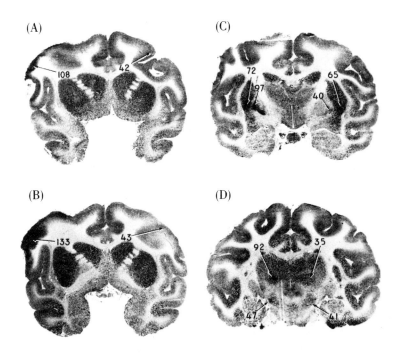

Local glucose utilization during focal seizure
(μmoles·100 g^{-1}·min^{-1}

FIGURE 5

Effects of focal seizures produced by local application of pen-icillin to motor cortex on local, cerebral glucose utilization in the rhesus monkey. The penicillin was applied to the hand and face area of the left motor cortex. The left side of the brain is on the left in each of the autoradiographs in the figure. The numbers are the rates of local cerebral glucose utilization in μ moles · 100 g tissue^{-1} · min^{-1}. (A) Motor cortex in region of penicillin application and corresponding region of contra-lateral motor cortex. (B) Ipsilateral and contralateral motor cortical regions remote from area of penicillin applications. (C) Ipsilateral and contralateral putamen and globus pallidus. (D) Ipsilateral and contralateral thalamic nuclei and sub-stantia nigra. (From Sokoloff, 1977.)

VISUAL OCCLUSION IN THE RAT   In the rat, the visual system is 80 to 85 percent crossed at the optic chiasma (Lashley, 1934; Montero and Guillery, 1968), and unilateral enucleation removes most of the visual input to the central visual structures of the contralateral side. In the conscious rat, studied 2–24 hours after unilateral enucleation, there are marked decrements in glucose utilization in the contralateral superior colliculus, lateral geniculate ganglion, and visual cortex as compared to the ipsilateral side (Kennedy et al., 1975).

VISUAL OCCLUSION IN THE MONKEY   In animals with binocular visual systems, such as the rhesus monkey, there is only approximately 50 percent crossing of the visual pathways; and the structures of the visual system on each side of the brain receive equal inputs from both retinae. Although each retina projects more or less equally to both hemispheres, their projections remain segregated and terminate in six well-defined laminae in the lateral geniculate ganglia, three each for the ipsilateral and contralateral eyes (Hubel and Wiesel, 1968, 1972; Wiesel et al., 1974; Rakic, 1976). This segregation is preserved in the optic radiations that project the monocular representations of the two eyes for any segment of the visual field to adjacent regions of Layer IV of the striate cortex (Hubel and Wiesel, 1968, 1972). The cells responding to the input of each monocular terminal zone are distributed transversely through the thickness of the striate cortex, an arrangement resulting in a mosaic of columns, 0.3–0.5 mm in width, alternately representing the monocular inputs of the two eyes. The nature and distribution of these ocular dominance columns have previously been characterized by electrophysiological techniques (Hubel and Wiesel, 1968), by Nauta degeneration methods (Hubel and Wiesel, 1972), and by autoradiographic visualization of axonal and transneuronal transport of [^{3}H]proline- and [^{3}H]fucose-labeled protein and/

or glycoprotein (Wiesel et al., 1974; Rakic, 1976). Bilateral or unilateral visual deprivation, either by enucleation or by the insertion of opaque plastic discs, produces consistent changes in the pattern of distribution of the rates of glucose consumption—all clearly visible in the autoradiographs—that coincide closely with the changes in functional activity expected from known physiological and anatomical properties of the binocular visual system (Kennedy et al., 1976).

In animals with intact binocular vision, no bilateral asymmetry is seen in the autoradiographs of the structures of the visual system (Figures 6A and 7A). The lateral geniculate ganglia and oculomotor nuclei appear to be of fairly uniform density and essentially the same on both sides (Figure 6A). The visual cortex is also the same on both sides (Figure 7A), but throughout all of Area 17 there is heterogeneous density distributed in a characteristic laminar pattern. These observations indicate that in animals with binocular visual input the rates of glucose consumption in the visual pathways are essentially equal on both sides of the brain and relatively uniform in the oculomotor nuclei and lateral geniculate ganglia but are markedly different in the various layers of the striate cortex.

Autoradiographs from animals with both eyes occluded exhibit generally decreased labeling of all components of the visual system, but the bilateral symmetry is fully retained (Figures 6B, and 7B), and the density within each lateral geniculate body is fairly uniform (Figure 6B). In the striate cortex, however, the marked differences in the densities of the various layers seen in the animals with intact bilateral vision (Figure 7A) are virtually absent so that, except for a faint delineation of a band within Layer IV, the concentration of the label is essentially homogeneous throughout the striate cortex (Figure 7B).

Autoradiographs from monkeys with only monocular input because of unilateral visual occlusion exhibit markedly dif-

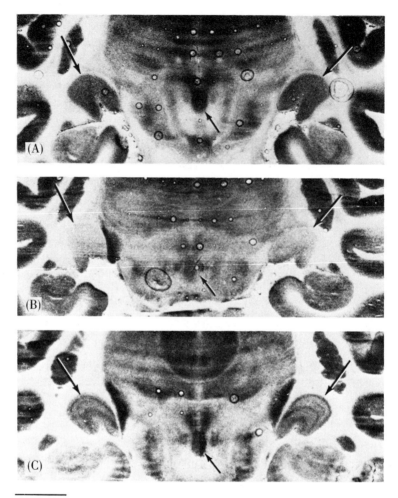

5.0 mm

ferent patterns from those described above. Both lateral geniculate bodies exhibit exactly inverse patterns of alternating
dark and light bands corresponding to the known laminae
representing the regions receiving the different inputs from
the retinae of the intact and occluded eyes (Figure 6C). Bilateral asymmetry is also seen in the oculomotor nuclear complex; a lower density is apparent in the nuclear complex contralateral to the occluded eye (Figure 6C). In the striate cortex,
the pattern of distribution of the $[^{14}C]DG$-6-P appears to be

FIGURE 6

Autoradiograph of coronal brain sections of monkey at the level of the lateral geniculate bodies. Long arrows point to the lateral geniculate bodies; short arrows point to oculomotor nuclear complex. (A) Animal with intact binocular vision. Note the bilateral symmetry and relative homogeneity of the lateral geniculate bodies and oculomotor nuclei. (B) Animal with bilateral visual occlusion. Note the reduced relative densities, the relative homogeneity, and the bilateral symmetry of the lateral geniculate bodies and oculomotor nuclei. (C) Animal with right eye occluded. The left side of the brain is on the left side of the photograph. Note the laminae and the inverse order of the dark and light bands in the two lateral geniculate bodies. Note also the lesser density of the oculomotor nuclear complex on the side contralateral to the occluded eye. (From Kennedy et al., 1976.)

a composite of the patterns seen in the animals with intact and bilaterally occluded visual input. The pattern found in the former regularly alternates with that of the latter in columns oriented perpendicularly to the cortical surface (Figure 7C). The dimensions, arrangement, and distribution of these columns are identical to those of the ocular dominance columns described by Hubel and Wiesel (Hubel and Wiesel, 1968, 1972; Wiesel et al., 1974). These columns reflect the interdigitation of the representations of the two retinae in the visual cortex. Each element in the visual fields is represented by a pair of contiguous bands in the visual cortex, one for each of the two retinae or their portions that correspond to the given point in the visual fields. With symmetrical visual input bilaterally, the columns representing the two eyes are equally active and, therefore, not visualized in the autoradiographs (Figure 7A). When one eye is blocked, however, only those columns representing the blocked eye become

FIGURE 7

Autoradiographs of coronal brain sections from rhesus monkeys at the level of the striate cortex. (A) Animal with normal binocular vision. Note the laminar distribution of the density; the dark band corresponds to Layer IV. (B) Animal with bilateral visual deprivation. Note the almost uniform and reduced relative density, especially the virtual disappearance of the dark band corresponding to Layer IV. (C) Animal with right eye occluded. The half-brain on the left side of the photograph represents the left hemisphere contralateral to the occluded eye. Note the alternate dark and light striations (each approximately 0.3–0.4 mm in width) that represent the ocular dominance columns. These columns are most apparent in the dark band corresponding to Layer IV, but they extend through the entire thickness of the cortex. The arrows point to regions of bilateral asymmetry where the ocular dominance columns are absent. These are presumably areas that normally receive only monocular input. The one on the left, contralateral to the occluded eye, has a continuous dark lamina corresponding to Layer IV, which is completely absent on the side ipsilateral to the occluded eye. These regions are believed to be the loci of the cortical representations of the blind spots. (From Kennedy et al., 1976.)

metabolically less active, and the autoradiographs then display the alternate bands of normal and depressed activities corresponding to the regions of visual cortical representation of the two eyes (Figure 7C).

There can be seen in the autoradiographs from the animals with unilateral visual deprivation a pair of regions in the folded calcarine cortex that exhibit bilateral asymmetry (Figure 7C). The ocular dominance columns are absent on both sides; but on the side contralateral to the occluded eye this region has the appearance of visual cortex from an animal with normal bilateral vision, and on the ipsilateral side this region looks

(A)

(B)

(C)

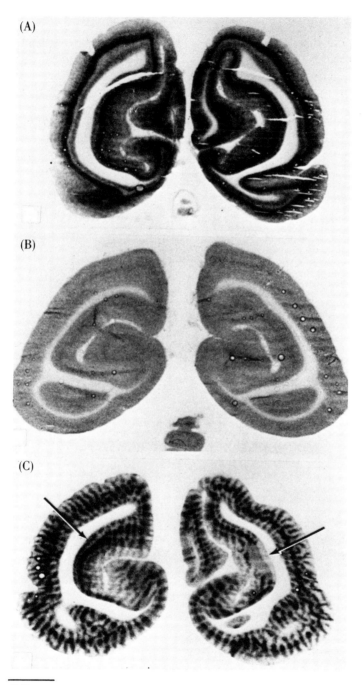

5.0 mm

like cortex from an animal with both eyes occluded (Figure 7). These regions are the loci of the cortical representation of the blind spots of the visual fields and normally have only monocular input (Kennedy et al., 1975, 1976). The area of the optic disc in the nasal half of each retina cannot transmit to this region of the contralateral striate cortex, which, therefore, receives its sole input from an area in the temporal half of the ipsilateral retina. Occlusion of one eye deprives this region of the ipsilateral striate cortex of all input, whereas the corresponding region of the contralateral striate cortex retains uninterrupted input from the intact eye. The metabolic reflection of this ipsilateral monocular input is seen in the autoradiograph in Figure 7C.

The results of these studies with the [^{14}C]deoxyglucose method in the binocular visual system of the monkey represent the most dramatic demonstration of the close relationship between physiological changes in functional activity and the rate of energy metabolism in specific components of the central nervous system.

APPLICATIONS OF
THE DEOXYGLUCOSE METHOD

The results of studies such as those described earlier on the effects of experimentally induced focal alterations of functional activity on local glucose utilization have demonstrated a close coupling between local functional activity and energy metabolism in the central nervous system. The effects are often so pronounced that they can be visualized directly on the autoradiographs, which provide pictorial representations of the relative rates of glucose utilization throughout the brain. This technique of autoradiographic visualization of evoked metabolic responses offers a powerful tool to map functional neural pathways simultaneously in all anatomical components of the central nervous system, and extensive use has been made of it for this purpose (Plum et al., 1976). The results have clearly demonstrated the effectiveness of metabolic responses, either positive or negative, in identifying regions of the central nervous system involved in specific functions.

The method has been used most extensively in qualitative studies in which regions of altered functional activity are identified by the change in their visual appearance relative to other regions in the autoradiographs. Such qualitative studies are effective only when the effects are lateralized to one side or when only a few discrete regions are affected; other regions serve as the controls. Quantitative comparisons cannot, however, be made for equivalent regions between two or more animals. To make quantitative comparisons between animals, the fully quantitative method must be used because it takes into account the various factors (particularly the plasma glucose level) that influence the magnitude of labeling of the tissues. The method must be used quantitatively when the experimental procedure produces systemic effects and alters metabolism in many regions of the brain.

A comprehensive review of the many qualitative and quantitative applications of the method is beyond the scope of this report. Only some of the many neurophysiological, neuroanatomical, pharmacological, and pathophysiological applications of the method in which our laboratory has been involved will be noted, to illustrate the broad extent of its potential usefulness.

Neurophysiological and neuroanatomical applications

Many of the physiological applications of the [^{14}C]deoxyglucose method were in studies designed to test the method and to examine the relationship between local cerebral functional and metabolic activities. These applications have been described in preceding sections. The most dramatic results have been obtained in the visual systems of the monkey and the rat. The method has, for example, been used to define the nature, conformation, and distribution of the ocular dominance columns in the striate cortex of the monkey (Figure 7C) (Kennedy et al., 1976). It has been used by Hubel et al. (1978) to do the same for the orientation columns in the striate cortex of the monkey. A by-product of the studies of the ocular dominance columns was the identification of the loci of the visual cortical representation of the blind spots of the visual fields (Figure 7C) (Kennedy et al., 1976). Studies are in progress to map the pathways of higher visual functions beyond the striate cortex; the results thus far demonstrate extensive areas of involvement of the inferior temporal cortex in visual processing (Macko et al., 1982). Des Rosiers et al. (1978) have used the method to demonstrate functional plasticity in the striate cortex of the infant monkey. The ocular dominance columns are already present on the first day of life; but when one eye is kept patched for 3 months, the columns representing the open eye broaden and completely take over the adjacent regions of cortex containing the columns for the eye that had been patched. Inasmuch as there is no longer any cortical representation for the patched eye, the animal becomes functionally blind in one eye. This phenomenon is almost certainly the basis for the cortical blindness or amblyopia that often occurs in children with uncorrected strabismus.

There have also been extensive studies of the visual system of the rat. This species has little if any binocular vision and, therefore, lacks the ocular dominance columns. Batipps et al. (1981) have compared the rates of local cerebral glucose utilization in albino and Norway brown rats during exposure to ambient light. The rates in the two strains were essentially the same throughout the brain except in the components of the primary visual system. The metabolic rates in the superior colliculus, lateral geniculate, and visual cortex of the albino rat were significantly lower than those in the pigmented rat. Miyaoka et al. (1979a) have studied the influence of the intensity of retinal stimulation with randomly spaced light flashes on the metabolic rates in the visual systems of the two strains. In dark-adapted animals, there is relatively little difference between the two strains. With increasing intensity of light, the rates of glucose utilization first increase in the primary projection areas of the retina (e.g., superficial layer of the superior colliculus and lateral geniculate body); and the slopes of the increase are steeper in the albino rat (Figure 8). At 7 lux, however, the metabolic rates peak in the albino rat and then decrease with increasing light intensity. In contrast, the metabolic rates in the pigmented rat rise until they reach a plateau at about 700 lux, approximately the ambient light intensity in the laboratory. At this level, the metabolic rates in the visual structures of the albino rat are considerably below those of the pigmented rat. These results are consistent with the greater intensity of light reaching the visual cells of the retina in the albino rats because of lack of pigment and the subsequent damage to the rods at higher light intensities. It is of considerable interest that the rates of glucose utilization in these visual structures obey the Weber-Fechner Law, that is, the metabolic rate is directly proportional to the logarithm of the intensity of stimulation (Miyaoka et al., 1979a). In-

FIGURE 8

Effects of intensity of retinal illumination with randomly spaced light flashes on local cerebral glucose utilization in components of the visual system of the albino and Norway brown rat. Note that the local glucose utilization is proportional to the logarithm of the intensity of illumination, at least at lower levels of intensity, in the primary projection areas of the retina. (From Miyaoka et al., 1979a.)

SUPERIOR COLLICULUS

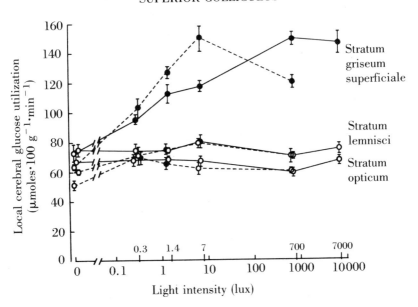

POSTEROLATERAL NUCLEUS OF THE THALAMUS

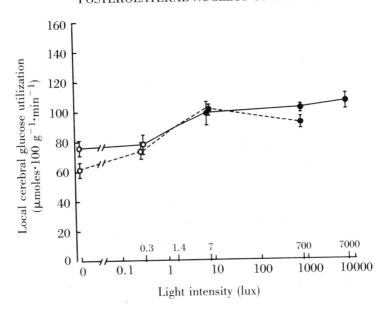

-- Albino
— Pigmented

LATERAL GENICULATE NUCLEUS

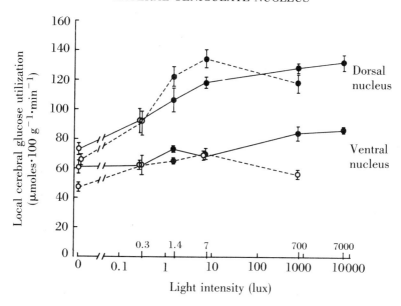

VISUAL CORTEX

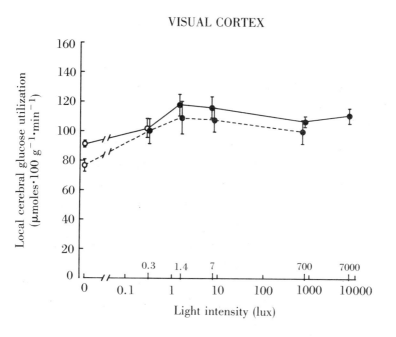

-- Albino
— Pigmented

asmuch as this law was first developed from behavioral man-
ifestations, these results imply that there is a quantitative
relationship between behavioral and metabolic responses.

Although less extensive, there have also been applications
of the method to other sensory systems. In studies of the olfactory system, Sharp et al. (1975) have found that olfactory
stimulation with specific odors activates glucose utilization
in localized regions of the olfactory bulb. In addition to the
experiments in the auditory system described earlier, there
have been studies of tonotopic representation in the auditory
system. Webster et al. (1978) have obtained clear evidence
of selective regions of metabolic activation in the cochlear
nucleus, superior olivary complex, nuclei of the lateral lem-
nisci, and the inferior colliculus in cats in response to different
frequencies of auditory stimulation. Similar results have been
obtained by Silverman et al. (1977) in the rat and the guinea
pig. Studies of the sensory cortex have demonstrated metabolic
activation of the "whisker barrels" by stimulation of the vi-
brissae in the rat (Durham and Woolsey, 1977; Hand et al.,
1978). Each vibrissa is represented in a discrete region of
the sensory cortex (Figure 9); their precise location and extent
have been elegantly mapped by Hand et al. (1978) and Hand
(1981) by means of the $[^{14}C]$deoxyglucose method.

Thus far there has been relatively little application of the
method to the physiology of motor functions. Kennedy et al.
(1980) have studied monkeys that were conditioned to perform
a task with one hand in response to visual cues; in the monkeys that were performing they observed metabolic activation
throughout the appropriate areas of the motor as well as sensory systems from the cortex to the spinal cord.

An interesting physiological application of the $[^{14}C]$de-
oxyglucose method has been to the study of circadian rhythms
in the central nervous system. Schwartz and his co-workers
(1977, 1980) found that the suprachiasmatic nucleus in the
rat exhibits circadian rhythmicity in metabolic activity, high

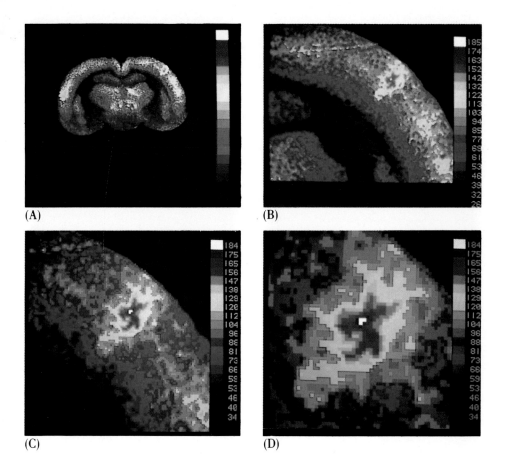

(A) (B) (C) (D)

FIGURE 9

Metabolic activation of "whisker barrel" in right sensory cortex of rat by stroking of single vibrissa on left side of face. (A) The entire section of the rat brain at the level of the "whisker barrel." (B–D) The results of rescanning at higher resolution and/or zooming. (D) A scan at highest resolution; each pixel is equivalent to 25 μm. The experiment was carried out by Hand et al. (1978), and the color-coded image processing was done by the method of Goochee et al. (1980). (From Hand, 1981; reproduced with permission of P. Hand.)

during the day and low during the night (Figure 10). None of the other structures in the brain that they examined showed rhythmic activity. The normally low activity present in the nucleus in the dark could be markedly increased by light, but darkness did not reduce the glucose utilization during the day. The rhythm is entrained to light; reversal of the light–dark cycle leads not only to reversal of the rhythm in running activity but also in the cycle of metabolic activity in the suprachiasmatic nucleus. These studies lend support to a role of the suprachiasmatic nucleus in the organization of circadian rhythms in the central nervous system.

Studies of circadian rhythms with the deoxyglucose method have been extended to natural sleep in monkeys. The results have demonstrated that during slow-wave, non-REM sleep, glucose utilization is diffusely depressed 25–30 percent throughout the central nervous system (Kennedy et al., 1982). No structure in the brain showed an increased rate of glucose utilization, not even structures proposed as hypnogenic centers, which allegedly are activated and depress functional activity in the other parts of the nervous system. The generalized uniformity of the metabolic depression suggests a chemical rather than a neurophysiological mechanism as the basis of slow-wave sleep. Studies in REM sleep have not yet been successfully carried out because of the short duration of REM episodes relative to the time-resolution of the method.

Much of our knowledge of neurophysiology has been derived from studies of the electrical activity of the nervous system. Indeed, from the heavy emphasis that has been placed on electrophysiology, one might gather that the brain is really an electric organ rather than a chemical one that functions mainly by the release of chemical transmitters at synapses. Nevertheless, electrical activity is unquestionably fundamental to the process of conduction, and it is appropriate to inquire how the local metabolic activities revealed by the [^{14}C]deoxyglucose method are related to the electrical activity

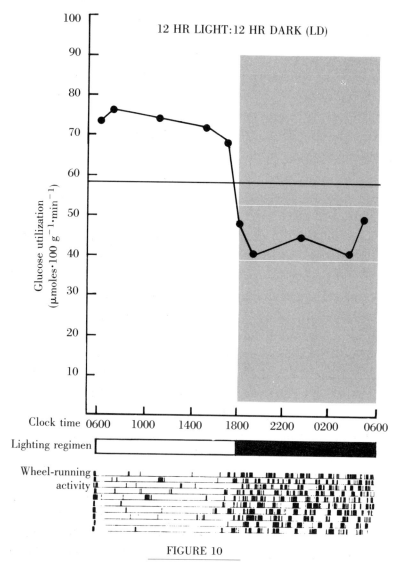

FIGURE 10

Circadian rhythm in glucose utilization in suprachiasmatic nucleus in the rat. (A) Animals entrained to 12 hours of light during day and 12 hours of darkness during night. (B) Animals entrained to opposite light–dark regimen. (From Schwartz et al., 1980.)

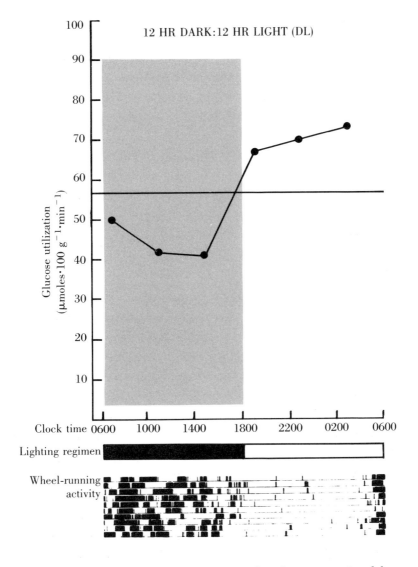

of the nervous system. This question has been examined by Yarowsky and his co-workers (1983) in the superior cervical ganglion of the rat. The advantage of this structure is that its preganglionic input and postganglionic ouput can be isolated and electrically stimulated and/or monitored in vivo. The results thus far indicate a clear relationship between electrical input to the ganglion and its metabolic activity. Glucose utilization in the superior cervical ganglion is enhanced by electrical stimulation of the afferent nerves. The metabolic activation is frequency dependent in the physiological range of 5 to 15 Hz, increasing in magnitude with increasing frequency of the stimulation (Figure 11). Similar effects of electrical stimulation on the oxygen and glucose consumptions of the excised ganglion studied in vitro have been observed (Larrabee, 1958; Horowicz and Larrabee, 1958; Friedli, 1978). Recent studies have also shown that antidromic stimulation of the postganglionic efferent pathways from the ganglion has similar effects; stimulation of the external carotid nerve antidromically activates glucose utilization in the region of distribution of the cell bodies of this efferent pathway, indicating that not only the preganglionic axonal terminals are metabolically activated, but the postganglionic cell bodies as well (Yarowsky et al., 1980). As already demonstrated in the neurohypophysis (Mata et al., 1980), the effects of electrical stimulation on energy metabolism in the superior cervical ganglion are also probably due to the ionic currents associated with the spike activity and the consequent activation of the Na^+, K^+-ATPase activity to restore the ionic gradients. Electrical stimulation of the afferents to sympathetic ganglia have been shown to increase extracellular K^+ concentration (Friedli, 1978; Galvan et al., 1979). Each spike is normally associated with a sharp transient rise in extracellular K^+ concentration, which then rapidly falls and transiently undershoots before returning to the normal level (Galvan et al., 1979); ouabain slows the decline in K^+ concentration after the spike and eliminates the undershoot. Continuous stimulation at a frequency of 6 Hz produces a sustained increase in cellular K^+ concentration (Galvan et al., 1979). It is likely that the increased extracellular K^+ concentration and, almost certainly, the increased intracellular Na^+ concentration activate the Na^+, K^+-ATPase, which in turn leads to the increased glucose utilization.

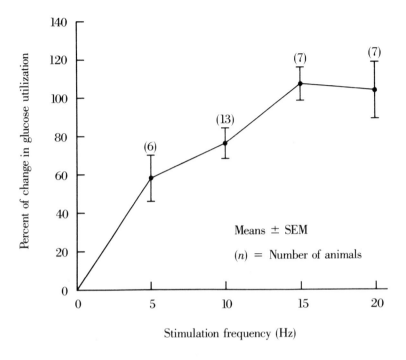

FIGURE 11

Relationship between frequency of electrical stimulation of cervical sympathetic trunk and the percentage of increase in the rate of glucose utilization in the superior cervical ganglion above that of the control ganglion on the other side. The values represent the means ± SEM of the individual percentage effects. (From Yarowsky et al., 1983.)

Neuroendocrinological applications

The deoxyglucose method has thus far been applied only sparingly to neuroendocrinology. Several studies have, however, been carried out or are in progress.

HYPOTHALAMONEUROHYPOPHYSIAL SYSTEM Physiological stimulation of the neurohypophysial system by salt-loading,

52

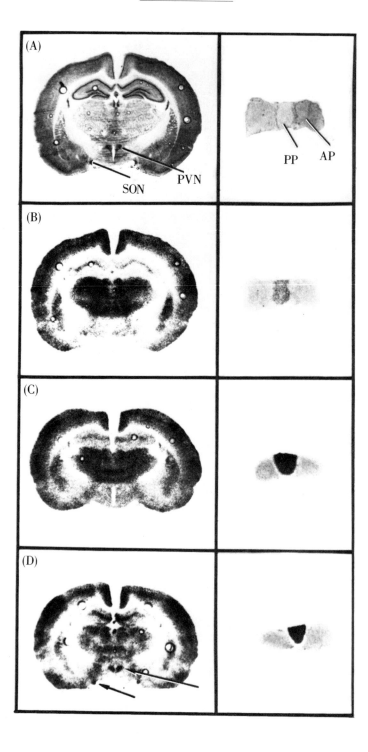

which enhances vasopressin secretion, has been found to be associated with increased glucose utilization in the posterior pituitary (Figure 12) (Schwartz et al., 1979). Surprisingly, there were no detectable effects in the supraoptic and paraventricular nuclei, the loci of the cell bodies that project to the posterior pituitary. Obviously the entire pathway had been activated by the osmotic stimulation. The posterior pituitary is composed (approximately 42 percent) of axonal terminals of the hypothalamohypophysial tract (Nordmann, 1977), and the discrepancy between the effects on the cell bodies and in the regions of termination of their projections may reflect a sensitivity of axonal terminals and/or synaptic elements greater than that of perikarya to metabolic activation. That the supraoptic and paraventricular nuclei can be metabolically

FIGURE 12

[^{14}C]Deoxyglucose autoradiographs (B–D) and stained histological sections (A) of coronal brain sections (left) and pituitary sections (right). (A) These photographs illustrate the positions of the supraoptic (SON) and paraventricular (PVN) nuclei in the brain section after cresyl violet (Nissl) staining. The positions of the posterior pituitary (PP) and anterior pituitary (AP) after toluidine blue staining are illustrated on the right side. (B) These autoradiographs are characteristic of control rats, which were allowed to drink water freely. (C) Autoradiographs of brain and pituitary typical of dehydrated rats, which were given 2 percent NaCl to drink for 5 days. Note the intense labeling in the posterior pituitary, without comparable change in the SON or PVN. (D) Autoradiographs characteristic of normal rats given an intravenous injection of an α-adrenergic blocker, phenoxybenzamine (20 mg/kg), approximately 45 to 60 minutes before injection of [^{14}C]deoxyglucose. Note the dramatic increase in labeling of the SON, PVN, and PP. (From Schwartz et al., 1979.)

activated is evident from the effects of the α-adrenergic blocking agent phenoxybenzamine (Figure 12), or of any other condition that produces hypotension (Savaki et al., 1982). In hypotension, however, these nuclei are activated by reflex activity, and it may well be that it is the afferent axonal terminals in these nuclei rather than the cell bodies that exhibit the increased utilization of glucose.

Kadekaro et al. (1983) have applied the deoxyglucose method to studies of the Brattleboro rat, a varient of the Long-Evans strain with a recessive genetic defect in vasopressin synthesis. This rat exhibits the characteristic signs of diabetes insipidus: an abnormally high water intake and a high output of hypoosmolar urine. Despite the deficiency in vasopressin synthesis, glucose utilization was found to be markedly increased in the posterior pituitary and also in the subfornical organ, a structure that has been found to mediate drinking behavior in response to high plasma levels of angiotensin II; and angiotensin II is elevated in the Brattleboro rat. As in the normal rat stimulated by salt-loading, the supraoptic and paraventricular nuclei were not metabolically more active in the Brattleboro rat. The reason for the high metabolic rate in the posterior pituitary of the Brattleboro rat is still obscure, but it may be related to histological changes that also are present in the gland.

ALTERED THYROID FUNCTION Hyperthyroidism is known not to alter the average energy metabolism in the mature brain as a whole (Sokoloff et al., 1953). Studies with the deoxyglucose method (D. Dow-Edwards and C.B. Smith, unpublished observations) reveal that there are also no changes in glucose utilization in any anatomical components of the mature brain of the rat. The thyroid hormones are, however, fundamentally involved in the structural and functional maturation of the brain (Eayrs, 1964). Dow-Edwards et al. (1982) have applied the deoxyglucose method to rats radiothyroidectomized at birth but studied at approximately 5 months of age when the brain of normal rats has achieved maturity. Glucose utilization was significantly reduced in all regions of the brain examined. Particularly affected were the cerebral cortical regions and the sensory systems, particularly the auditory system. These metabolic changes are consistent with the histological pattern of impaired brain development in cretinism.

SEX HORMONES AND SEXUAL BEHAVIOR   The deoxyglucose method has been used to demonstrate selective metabolic activation of a number of structures in the female rat brain by vaginocervical stimulation that also elicited lordotic behavior (Allen et al., 1981). The structures so activated were the medial preoptic nucleus, mesencephalic reticular formation, red nucleus of the stria terminalis, dorsal raphe nucleus, and the globus pallidus. Some, but not all, of these areas had been previously shown by electrical recording, placement of lesions, and stimulating techniques to participate in the behavioral and physiological responses to coitus. These results provide additional information about the concurrent processing of sensory stimulation in the brain and also indicate that the medial preoptic area is a receptive area for copulatory stimulation.

The female gonadal hormones—estrogen and progesterone—influence sexual behavior and have potent influences on the central nervous system, particularly the hypothalamus. Porrino et al. (1982) have used the deoxyglucose method in an attempt to identify regions of the brain affected by these hormones. In ovariectomized rats, estradiol treatment stimulated glucose utilization in anterior, ventromedial, lateral, and posterior areas of the hypothalamus. Progesterone alone had very little effect in these areas. When progesterone was administered to animals that had been implanted previously with estradiol, glucose utilization was reduced in the lateral preoptic area, medial preoptic area, and anterior hypothalamus below the levels of the ovariectomized controls. These data suggest an anatomical separation of the effects of gonadal steroids in the hypothalamus; estradiol may facilitate neural activity in the mid and posterior areas of the hypothalamus, whereas estrogen in combination with progesterone suppresses activity in the anterior preoptic area. These two patterns in the anterior and posterior hypothalamus may reflect differential involvement in feminine sexual behavior.

Neuropharmacological applications

The ability of the deoxyglucose method to map the entire brain for localized regions of altered functional activity on the basis of changes in energy metabolism offers a potent tool to identify the neural sites of action of agents with neuropharmacological and psychopharmacological actions. It does not, however, discriminate between the direct and indirect effects of the drug. An entire pathway may be activated even though the direct action of the drug may be exerted only at the origin of the pathway. This is useful for relating behavioral effects to central actions, but it is a disadvantage if the goal is to identify the primary site of action of the drug. To discriminate between direct and indirect actions of a drug, the [^{14}C]deoxyglucose method must be combined with selectively placed lesions in the central nervous system that interrupt afferent pathways to the structure in question. If the metabolic effect of the drug then remains, it is due to direct action; if lost, the effect is likely to be indirect and mediated via the interrupted pathway. Nevertheless, the method has proved to be useful in a number of pharmacological studies.

EFFECTS OF γ-BUTYROLACTONE γ-Hydroxybutyrate and γ-butyrolactone, which is hydrolyzed to γ-hydroxybutyrate in plasma, produce trancelike behavioral states associated with marked suppression of electroencephalographic activity (Roth and Giarman, 1966). These effects are reversible, and these drugs have been used clinically as anesthetic adjuvants. There is evidence that these agents lower neuronal activity in the nigrostriatal pathway and may act by inhibition of dopaminergic synapses (Roth, 1976). Studies in rats with the [^{14}C]deoxyglucose technique have demonstrated that γ-butyrolactone produces profound dose-dependent reductions of glucose utilization throughout the brain (Wolfson et al., 1977). At the highest doses studied, 600 mg/kg of body weight, glucose utilization was reduced by approximately 75 percent in gray matter and 33 percent in white matter, but there was no obvious further specificity with respect to the local cerebral structures affected. The reversibility of the effects and the magnitude and diffuseness of the depression of cerebral metabolic rate suggests that this drug might be considered as a chemical substitute for hypothermia in conditions in which profound reversible reduction of cerebral metabolism is desired.

EFFECTS OF D-LYSERGIC ACID DIETHYLAMIDE   The effects of the potent psychotomimetic agent D-lysergic acid diethylamide have been examined in the rat (Shinohara et al., 1976). In doses of 12.5 to 125 μg/kg, it caused dose-dependent reductions in glucose utilization in a number of cerebral structures. With increasing dosage, more structures were affected and to a greater degree. There was no pattern in the distribution of the effects, at least none discernible at the present level of resolution that might contribute to the understanding of the drug's psychotomimetic actions.

EFFECTS OF MORPHINE ADDICTION AND WITHDRAWAL Acute morphine administration depresses glucose utilization in many areas of the brain, but the specific effects of morphine could not be distinguished from those of the hypercapnia produced by the associated respiratory depression (Sakurada et al., 1976). In contrast, morphine addiction, produced within 24 hours by a single subcutaneous injection of 150 mg/kg of morphine base in an oil emulsion, reduces glucose utilization in a large number of gray structures in the absence of changes in arterial pCO_2. White matter appears to be unaffected. Naloxone (1 mg/kg subcutaneously) reduces glucose utilization in a number of structures when administered to normal rats; but when given to the morphine-addicted animals it produces an acute withdrawal syndrome and reverses the reductions of glucose utilization in several structures, most strikingly in the habenula (Sakurada et al., 1976).

PHARMACOLOGICAL STUDIES OF DOPAMINERGIC SYSTEMS
The most extensive applications of the deoxyglucose method to pharmacology have been in studies of dopaminergic systems. Ascending dopaminergic pathways appear to have a potent influence on glucose utilization in the forebrain of rats. Electrolytic lesions placed unilaterally in the lateral hypothalamus or pars compacta of the substantia nigra caused marked ipsilateral reductions of glucose metabolism in numerous forebrain structures rostral to the lesion, particularly the frontal cerebral cortex, caudate-putamen, and parts of the thalamus (Schwartz et al., 1976; Schwartz, 1978). Similar lesions in the locus coeruleus had no such effects.

Enhancement of dopaminergic synaptic activity by administration of apomorphine (an agonist of dopamine) (Brown and Wolfson, 1978) or of amphetamine (Wechsler et al., 1979), which stimulates release of dopamine at the synapse, produces marked increases in glucose consumption in some of the components of the extrapyramidal system known or suspected to contain dopamine-receptive cells. With both drugs, the greatest increases noted were in the zona reticulata of the substantia nigra and the subthalamic nucleus. Sur-

prisingly, none of the components of the dopaminergic me-
solimbic system appeared to be affected.

The studies with amphetamine (Wechsler et al., 1979)
were carried out with the fully quantitative [^{14}C]deoxyglucose
method. The results in Table 5 illustrate the comprehen-
siveness with which this method surveys the entire brain for
sites of altered activity due to actions of the drug. It also
allows for quantitative comparison of the relative potencies
of related drugs. For example, in Table 5, the comparative
effects of d-amphetamine and the less potent dopaminergic
agent l-amphetamine are compared; the quantitative results
clearly reveal that the effects of l-amphetamine on local cerebral glucose utilization are more limited in distribution and
of lesser magnitude than are those of d-amphetamine. Indeed,
in similar quantitative studies with apomorphine, McCulloch
et al. (1979, 1980a) have been able to generate complete
dose-response curves for the effects of the drug on the rates
of glucose utilization in various components of dopaminergic
systems. They have also demonstrated metabolically the development of supersensitivity to apomorphine in rats maintained chronically on the dopamine antagonist haloperidol (J.
McCulloch, H.E. Savaki, A. Pert, W. Bunney, and L. Sokoloff, unpublished observations). In the course of these
studies with apomorphine, McCulloch et al. (1980b) obtained
evidence of a retinal dopaminergic system that projects spe-
cifically to the superficial layer of the superior colliculus in
the rat. Apomorphine administration activated metabolism in
the superficial layer of the superior colliculus, as well as in
other structures, but the effect in the superficial layer was
prevented by prior enucleation (Figure 13). Miyaoka (unpublished observations) subsequently observed that intraocu-
lar administration of minute amounts of apomorphine caused
increased glucose utilization only in the superficial layer of
the superior colliculus of the contralateral side.

TABLE 5

EFFECTS OF d-AMPHETAMINE
AND l-AMPHETAMINE ON
LOCAL CEREBRAL GLUCOSE UTILIZATION
IN THE CONSCIOUS RAT[a]

Structure	Control	d-Amphetamine	l-Amphetamine
GRAY MATTER			
Visual cortex	102 ± 8	135 ± 11[b]	105 ± 8
Auditory cortex	160 ± 11	162 ± 6	141 ± 6
Parietal cortex	109 ± 9	125 ± 10	116 ± 4
Sensory-motor cortex	118 ± 8	139 ± 9	111 ± 4
Olfactory cortex	100 ± 6	93 ± 5	94 ± 3
Frontal cortex	109 ± 10	130 ± 8	105 ± 4
Prefrontal cortex	146 ± 10	166 ± 7	154 ± 4
Thalamus			
Lateral nucleus	97 ± 5	114 ± 8	117 ± 6
Ventral nucleus	85 ± 7	108 ± 6[b]	96 ± 4
Habenula	118 ± 10	71 ± 5[c]	82 ± 2[c]
Dorsomedial nucleus	92 ± 6	111 ± 8	106 ± 6
Medial geniculate	116 ± 5	119 ± 4	116 ± 4
Lateral geniculate	79 ± 5	88 ± 5	84 ± 4
Hypothalamus	54 ± 5	56 ± 3	52 ± 3
Suprachiasmatic nucleus	94 ± 4	75 ± 4[c]	67 ± 1[c]
Mamillary body	117 ± 8	134 ± 5	142 ± 5[b]
Lateral olfactory nucleus[d]	92 ± 6	95 ± 5	99 ± 6
A$_{13}$	71 ± 4	91 ± 4[c]	81 ± 4
Hippocampus			
Ammon's horn	79 ± 5	73 ± 2	81 ± 6
Dentate gyrus	60 ± 4	55 ± 3	67 ± 7
Amygdala	46 ± 3	46 ± 3	44 ± 2
Septal nucleus	56 ± 3	55 ± 2	54 ± 3
Caudate nucleus	109 ± 5	132 ± 8[b]	127 ± 3[b]

TABLE 5 (Continued)

Structure	Control	d-Amphetamine	l-Amphetamine
GRAY MATTER			
Nucleus accumbens	76 ± 5	80 ± 3	78 ± 3
Globus pallidus	53 ± 3	64 ± 2[b]	65 ± 3[b]
Subthalamic nucleus	89 ± 6	149 ± 10[c]	107 ± 2
Substantia nigra			
Zona reticulata	58 ± 2	105 ± 4[c]	72 ± 4
Zona compacta	65 ± 4	88 ± 6[c]	72 ± 3
Red nucleus	76 ± 5	94 ± 5[b]	86 ± 2
Vestibular nucleus	121 ± 11	137 ± 5	130 ± 4
Cochlear nucleus	139 ± 6	126 ± 1	141 ± 5
Superior olivary nucleus	144 ± 4	143 ± 4	147 ± 6
Lateral lemniscus	107 ± 3	96 ± 5	98 ± 3
Inferior colliculus	193 ± 10	169 ± 5	150 ± 8[c]
Dorsal tegmental nucleus	109 ± 5	112 ± 7	122 ± 6
Superior colliculus	80 ± 5	89 ± 3	91 ± 3
Pontine gray	58 ± 4	65 ± 3	60 ± 1
Cerebellar flocculus	124 ± 10	146 ± 15	153 ± 10
Cerebellar hemispheres	55 ± 3	68 ± 6	64 ± 2
Cerebellar nuclei	102 ± 4	105 ± 8	110 ± 3
WHITE MATTER			
Corpus callosum	23 ± 3	24 ± 2	23 ± 1
Genu of corpus callosum	29 ± 2	30 ± 2	26 ± 2
Internal capsule	21 ± 1	24 ± 2	19 ± 2
Cerebellar white	28 ± 1	31 ± 2	31 ± 2

Source: Wechsler et al. (1979).

[a] All values are the means ± standard error of the mean for five animals.

[b] Significant difference from the control at the $p < 0.05$ level.

[c] Significant difference from the control at the $p < 0.01$ level.

[d] It was not possible to correlate precisely this area on autoradiographs with a specific structure in the rat brain. It is, however, most likely the lateral olfactory nucleus.

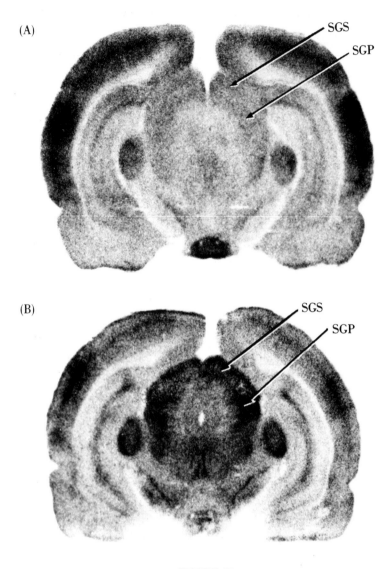

FIGURE 13

Representative autoradiographs at the level of the superior col-
liculus in dark-adapted rats studied in the dark. SGS, Stratum
griseum superficiale; SGP, stratum griseum profundum. (A)
Saline, intact visual system. (B) Apomorphine (1.5 mg/kg),
intact visual system. Note bilaterally increased optical density
(i.e., elevated glucose utilization) in both superficial and deep
laminae of the superior colliculus. (C) Saline, right eye en-

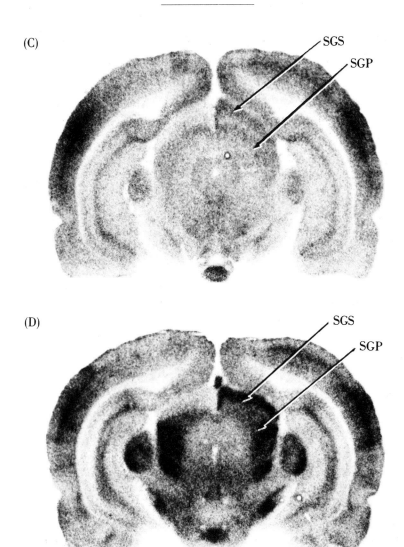

(C)

SGS

SGP

(D)

SGS

SGP

*ucleated. Asymmetrical optical density with reduction on con-
tralateral side is apparent within the superficial layer, whereas
in the deeper layer optical density remains symmetrical. (D)
Apomorphine (1.5 mg/kg), right eye enucleated. Note increased
optical density bilaterally in the deeper layer but only in the
right or ipsilateral superficial layer of the superior colliculus.
(From McCulloch et al., 1980b.)*

EFFECTS OF α- AND β-ADRENERGIC BLOCKING AGENTS
Savaki et al. (1978) have studied the effects of the α-adrenergic blocking agent phentolamine and the β-adrenergic
blocking agent propranolol. Both drugs produced widespread,
dose-dependent depressions of glucose utilization throughout
the brain but exhibit particularly striking and opposite effects
in the complete auditory pathway from the cochlear nucleus
to the auditory cortex. Propranolol markedly depressed and
phentolamine markedly enhanced glucose utilization in this
pathway. The functional significance of these effects is unknown, but they seem to correlate with corresponding effects
on the electrophysiological responsiveness of this sensory
system. Propranolol depresses and phentolamine enhances
the amplitude of all component of evoked auditory responses
(T. Furlow et al., 1980).

Neuropathophysiological applications

The application of the deoxyglucose method to the study of
pathological states has been limited because of uncertainties
about the values for the lumped and rate constants to be used.
There are, however, pathophysiological states in which there
is no structural damage to the tissue and in which the standard
values of the constants can be used. Several of these conditions have been and are continuing to be studied by the
[14C]deoxyglucose technique, both qualitatively and quantitatively.

CONVULSIVE STATES   The local injection of penicillin into
the motor cortex produces focal seizures manifested in specific
regions of the body contralaterally. The [14C]deoxyglucose
method has been used to map the spread of seizure activity
within the brain and to identify the structures with altered
functional activity during the seizure. The partial results of
one such experiment in the monkey are illustrated in Figure

5. Discrete regions of markedly increased glucose utilization are observed ipsilaterally in the motor cortex, basal ganglia, particularly the globus pallidus, thalamic nuclei, and contralaterally in the cerebellar cortex (Kennedy et al., 1975). Kato et al. (1980), Caveness et al. (1980), Hosokawa et al. (1980), and Caveness (1980) have carried out the most extensive studies of the propagation of the seizure activity in newborn and pubescent monkeys. The results indicate that the brain of the newborn monkey exhibits similar increases of glucose utilization in specific structures, but the pattern of distribution of the effects is less well defined than in pubescent monkeys. Collins et al. (1976) have carried out similar studies in the rat, with similar results; but they also obtained evidence on the basis of a local stimulation of glucose utilization of a "mirror focus" in the motor cortex contralateral to the side with the penicillin-induced epileptogenic focus.

Engel et al. (1978) have used the $[^{14}C]$deoxyglucose method to study seizures kindled in rats by daily electroconvulsive shocks. After a period of such treatment, the animals exhibit spontaneous seizures. Their results show marked increases in the limbic system, particularly the amygdala. The daily administration of the local anesthetic lidocaine kindles similar seizures in rats; Post et al. (1979) have obtained similar results in such seizures with particularly pronounced increases in glucose utilization in the amygdala, the hippocampus, and the enterorhinal cortex.

SPREADING CORTICAL DEPRESSION Shinohara et al. (1979) studied the effects of local applications of KCl on the dura overlying the parietal cortex of conscious rats or directly on the pial surface of the parietal cortex of anesthetized rats in order to determine whether K^+ stimulates cerebral energy metabolism in vivo, as it is well known to do in vitro. The results demonstrate a marked increase in cerebral cortical glucose utilization in response to the application of KCl; NaCl has no such effect (Figure 14). Such application of KCl, however, also produces the phenomenon of spreading cortical depression. This condition is characterized by a spread of transient, intense neuronal activity followed by membrane depolarization, electrical depression, and a negative shift in the cortical DC potential in all directions from the site of initiation at a rate of 2–5 mm/min. The depressed cortex also exhibits a number of chemical changes, including an increase in extracellular K^+, lost presumably from the cells. At the same time, when the cortical glucose utilization is increased, most subcortical structures that are functionally connected to the depressed cortex exhibit decreased rates of glucose utilization. During recovery from the spreading cortical depression, the glucose utilization in the cortex is still increased, but it is distributed in columns oriented perpendicularly through the cortex. This columnar arrangement may reflect the columnar functional and morphological arrangement of the cerebral cortex. It is likely that the increased glucose utilization in the cortex during spreading cortical depression is the consequence of the increased extracellular K^+ and activation of the NA^+, K^+-ATPase.

OPENING OF BLOOD–BRAIN BARRIER   Unilateral opening of the blood–brain barrier in rats by unilateral carotid injection with a hyperosmotic mannitol solution leads to widely distributed discrete regions of intensely increased glucose utilization in the ipsilateral hemisphere (Pappius et al., 1979). These focal regions of hypermetabolism may reflect local regions of seizure activity. The prior administration of diazepam prevents in most cases the appearance of these areas of increased metabolism (Pappius et al., 1979), and electroencephalographic recordings under similar experimental conditions reveal evidence of seizure activity (C. Fieschi et al., 1980).

FIGURE 14 ▶

Autoradiographs of sections of rat brains during spreading cortical depression and during recovery. The autoradiographs are pictorial representations of the relative rates of glucose utilization in various parts of the brain: the greater the density, the greater the rate of glucose utilization. The left sides of the brain are represented by the left hemispheres in the autoradiographs. In all the experiments illustrated, the control hemisphere was treated in the same way as the experimental side except that equivalent concentrations of NaCl rather than KCl were used. The NaCl did not lead to any detectable differences from hemispheres over which the skull was left intact and to which no NaCl was applied. (A) Autoradiographs of sections of brain at different levels of cerebral cortex from a conscious rat during spreading cortical depression induced on the left side by application of 5 M KCl to the intact dura overlying the left parietal cortex. The spreading depression was sustained by repeated applications of the KCl at 15- to 20-minute intervals throughout the experimental period. (B) Autoradiographs from sections of brain at the level of the parietal cortex from three animals under barbiturate anesthesia. The top section is from a normal anesthetized animal; the middle section is from an animal during unilateral spreading cortical depression induced and sustained by repeated applications of 80 mM KCl in artificial cerebrospinal fluid directly on the surface of the left parietooccipital cortex. At the bottom is a comparable section from an animal studied immediately after the return of cortical DC potential to normal after a single wave of spreading depression induced by a single application of 80 mM KCl to the parietooccipital cortex of the left side. (From Shinohara et al., 1979.)

(A)

Frontal cortex

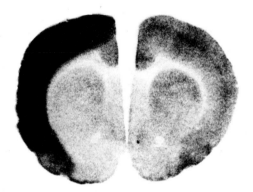

Sensory-motor cortex

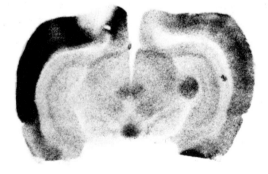

Parietal cortex

(B)

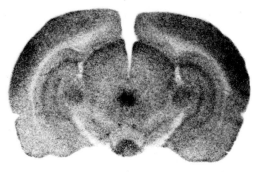

Control

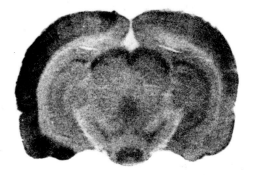

During spreading depression

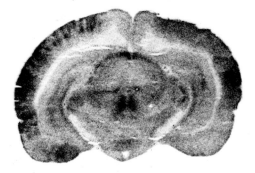

During recovery from
spreading depression

HYPOXEMIA   Pulsinelli and Duffy (1979) have studied the effects of controlled hypoxemia on local cerebral glucose utilization by means of the qualitative [^{14}C]deoxyglucose method. Hypoxemia was achieved by artificial ventilation of the animals with a mixture of N_2, N_2O, and O_2, adjusted to maintain the arterial pO_2 between 28 and 32 mm Hg. All the animals had had one common carotid artery ligated to limit the increases in cerebral blood flow and the amount of O_2 delivered to the brain. Their autoradiographs provide striking evidence of marked and disparate changes in glucose utilization in the various structural components of the brain. The hemisphere ipsilateral to the carotid ligation was not unexpectedly more severely affected. The most striking effects were markedly higher increases in glucose utilization in white matter than in gray matter (presumably due to the Pasteur effect) and the appearance of transverse cortical columns of high activity alternating with columns of low activity. In experiments with black plastic microspheres, they were able to show that the cortical columns were anatomically related to penetrating cortical arteries, with the columns of high metabolic activity lying between the arteries.

Miyaoka et al. (1979b) have also studied the effects of moderate hypoxemia in normal, spontaneously breathing, conscious rats without carotid ligation. The hypoxemia was produced by lowering the O_2 in the inspired air to approximately 7 percent. Although this procedure reduced arterial pO_2 to approximately 30 mm Hg, the cerebral hypoxia was probably less than in the studies of Pulsinelli and Duffy (1979) because of the intact cerebral circulation. The animals remained fully conscious under these experimental conditions, although they appeared subdued and less active. The quantitative [^{14}C]deoxyglucose method was employed, and rates of glucose utilization were determined. The results revealed many similarities to those of Pulsinelli and Duffy (1979).

There was a complete redistribution of the local rates of glucose utilization from the normal pattern. Metabolism in white matter was markedly increased. Many areas showed decreased rates of metabolism. Columns were seen in the cerebral cortex, and the caudate nucleus exhibited a strange lacelike heterogeneity quite distinct from its normal homogeneity. Despite the widespread changes, however, overall average glucose utilization remained unchanged. These results are of relevance to the studies by Kety and Schmidt (1948b), who found in humans that the breathing of 10 percent O_2 produced a wide variety of mental symptoms without altering the average O_2 consumption of the brain as a whole. The mental symptoms were probably the result of metabolic and functional changes in specific regions of the brain detectable only by methods like the deoxyglucose method that measure metabolic rate in the structural components of the brain.

NORMAL AGING   Although, strictly speaking, aging is not a pathophysiological condition, many of its behavioral consequences are directly attributable to decrements in functions of the central nervous system (Birren et al., 1963). Normal human aging has been found to be associated with a decrease in average glucose utilization of the brain as a whole (Sokoloff, 1966). Smith et al. (1980) have employed the quantitative [^{14}C]deoxyglucose method to study normal aging in Sprague-Dawley rats between 5–6 and 36 months of age. Their results show widespread, but not homogeneous, reductions of local cerebral glucose utilization with age. The sensory systems, particularly auditory and visual, are severely affected. The caudate nucleus is metabolically depressed, and preliminary experiments indicate that it loses responsivity to dopamine agonists, such as apomorphine, with age (C. Smith and J. McCulloch, unpublished observations). A striking effect was the loss of metabolically active neuropil in the cerebral cortex;

Layer IV is markedly decreased in metabolic activity and extent. Some of these changes may be related to specific functional disabilities that develop in old age.

MICROSCOPIC RESOLUTION

The resolution of the present [^{14}C]deoxyglucose method is at best 100–200 μm (Smith, 1983). The use of [^{3}H]deoxyglucose does not greatly improve the resolution when the standard autoradiographic procedure is used. The limiting factor is the diffusion and migration of the water-soluble, labeled compound in the tissue during the freezing of the brain and the cutting of the brain sections (Smith, 1983). Des Rosiers and Descarries (1978) attempted to extend the resolution of the method to the light and electron microscopic levels by the use of [^{3}H]deoxyglucose and dipping emulsion techniques applied to brain that was fixed, dehydrated, and embedded by perfusion in situ. They could localize grain counts over individual cells or portions of them, but loss of label and, therefore, also of quantitative reliability undoubtedly occurred. An alternative promising approach to microscopic resolution is the use of freeze-substitution techniques (Ornberg et al., 1979; Sejnowski et al., 1980; Smith, 1983).

THE [^{18}F]FLUORODEOXYGLUCOSE TECHNIQUE

Because the deoxyglucose method requires the measurement of local concentrations of radioactivity in the individual components of the brain, it cannot be applied as originally designed to man. Recent developments in computerized emission tomography, however, have made it possible to measure local concentrations of labeled compounds in vivo in man. Emission tomography requires the use of γ radiation, preferably annihilation γ rays derived from positron emission. A positron-emitting derivative of deoxyglucose (2-[^{18}F]fluoro-2-deoxy-D-glucose) has been synthesized and found to retain the necessary biochemical properties of 2-deoxyglucose (Reivich et al., 1979). The method has, therefore, been adapted for use in humans with [^{18}F]fluorodeoxyglucose and positron-emission tomography (Reivich et al., 1979; Phelps et al., 1979).

The resolution of the method is still relatively limited (approximately 0.8–1.0 cm), but it is already proving to be useful in studies of the human visual (Figure 15) and auditory systems (Phelps et al., 1981) and of clinical conditions, such as focal epilepsy (Kuhl et al., 1979, 1980), Huntington's disease (Kuhl et al., 1982), aging and dementia (Kuhl et al., 1983; Foster et al., 1983), and cerebral gliomas (DiChiro et al., 1983). This technique should prove to be immensely useful in studies of human local cerebral energy metabolism in normal states and in neurological and psychiatric disorders.

SUMMARY

The deoxyglucose method makes it possible to determine quantitatively the rates of glucose utilization simultaneously in all structural and functional components of the central nervous system and to display them pictorially superimposed on the anatomical structures in which they occur. Because of the close relationship between local functional activity and energy metabolism, the method makes it possible to identify all structures with increased or decreased functional activity in various physiological, pharmacological, and pathophysiological states. The images provided by the method do resemble histological sections of nervous tissue, and the method is, therefore, sometimes misconstrued to be a neuroanatomical method and is contrasted with physiological methods, such

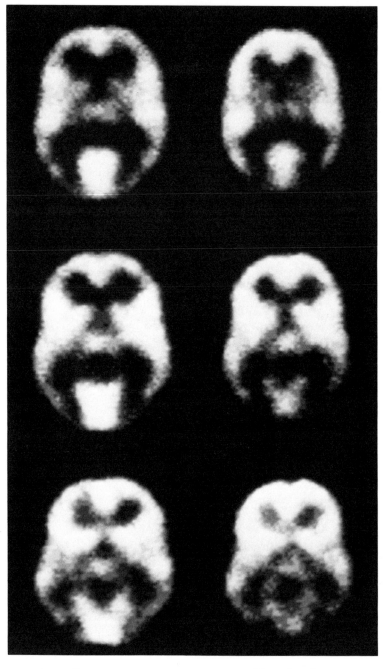

Eyes open   Eyes closed

FIGURE 15

Influence of visual input on glucose utilization of human cerebral cortex. Left column: *Three horizontal sections of the brain of a normal, conscious human male studied with [^{18}F]fluorodeoxyglucose technique while eyes were open.* Right column: *Same three sections studied with eyes closed. In each section, frontal is at the top and occipital at the bottom. The brighter the image, the higher the rate of glucose utilization; the darker the area, the lower the rate of glucose utilization. Note the decreased glucose utilization in the occipital cortex and the increased metabolic activity in the frontal cortex when the eyes are closed. (From Phelps et al., 1981.)*

as electrophysiological recording. This classification obscures the most significant and unique feature of the method. The images are not of structure but of a dynamic biochemical process—glucose utilization—that is as physiological as electrical activity. In most situations, changes in functional activity result in changes in energy metabolism, and the images can be used to visualize and identify the sites of altered activity. The images are, therefore, analogous to infrared maps; they record quantitatively the rates of a kinetic process and display them pictorially exactly where they exist. The fact that they depict the anatomical structures is fortuitous; it indicates that the rates of glucose utilization are distributed according to structure, and that specific functions in the nervous system are associated with specific anatomical structures. The deoxyglucose method represents, therefore, in a real sense, a new type of encephalography—metabolic encephalography. At the very least, it should serve as a valuable supplement to more conventional types, such as electroencephalography. Because, however, it provides a new means to examine another aspect of function simultaneously in all parts of the brain, it is hoped that it and its derivative, the

[^{18}F]fluorodeoxyglucose technique, will open new roads to the understanding of how the brain works in health and disease in animals and in man.

REFERENCES

Abrams, R., Ito, M., Frisinger, J.E., Patlak, C.S., Pettigrew, K.D., and Kennedy, C. (1984) Local cerebral glucose utilization in fetal and neonatal sheep. *Am. J. Physiol.*, in press.

Allen, T.O., Adler, N.T., Greenberg, J.H., and Reivich, M. (1981) Vaginocervical stimulation selectively increases metabolic activity in the rat brain. *Science* 211, 1070–1072.

Bachelard, H.S. (1971) Specificity and kinetic properties of monosaccharide uptake into guinea pig cerebral cortex *in vitro*. *J. Neurochem.* 18, 213–222.

Batipps, M., Miyaoka, M., Shinohara, M., Sokoloff, L., and Kennedy, C. (1981) Comparative rates of local cerebral glucose utilization in the visual system of conscious albino and pigmented rat. *Neurology* 31, 58–62.

Bidder, T.G. (1968) Hexose translocation across the blood–brain interface: Configurational aspects. *J. Neurochem.* 15, 867–874.

Birren, J.E., Butler, R.N., Greenhouse, S.W., Sokoloff, L., and Yarrow, M.R. (eds.) (1963) *Human Aging. A Biological and*

Behavioral Study. Public Health Service Publication No. 986, U.S. Government Printing Office, Washington, D.C.

Brown, L.L. and Wolfson, L.I. (1978) Apomorphine increases glucose utilization in the substantia nigra, subthalamic nucleus and corpus striatum of rat. *Brain Res.* 140, 188–193.

Caveness, W.F. (1969) Ontogeny of focal seizures. In H.H. Jasper, A.A. Ward, and A. Pope (eds.), *Basic Mechanisms of the Epilepsies*. Little, Brown and Co., Boston, pp. 517–534.

Caveness, W.F. (1980) Appendix: Tables of local cerebral glucose utilization in various experimental preparations. *Ann. Neurol.* 7, 230–237.

Caveness, W.F., Kato, M., Malamut, B.L., Hosokawa, S., Wakisaka, S., and O'Neill, R.R. (1980) Propagation of focal motor seizures in the pubescent monkey. *Ann. Neurol.* 7, 213–221.

Collins, R.C., Kennedy, C., Sokoloff, L., and Plum, F. (1976) Metabolic anatomy of focal motor seizures. *Arch. Neurol.* 33, 536–542.

Des Rosiers, M.H. and Descarries, L. (1978) Adaptation de la méthode au désoxyglucose a l'echelle cellulaire: Préparation histologique du système nerveux central en vue de la radioautographie à haute résolution. *C. R. Acad. Sci. Paris, Series D* 287, 153–156.

Des Rosiers, M.H., Sakurada, O., Jehle, J., Shinohara, M., Kennedy, C., and Sokoloff, L. (1978) Functional plasticity in the immature striate cortex of the monkey shown by the [^{14}C]deoxyglucose method. *Science* 200, 447–449.

DiChiro, G., Brooks, R.A., Sokoloff, L., Patronas, N.J., DeLaPaz, R.L., Smith, B.H., and Kornblith, P.L. (1983) Glycolytic rate and histologic grade of human cerebral gliomas: A study with [^{18}F]fluorodeoxyglucose and positron emission tomography. In E.-D. Heiss and M.E. Phelps (eds.), *Positron Emission Tomography of the Brain*. Springer-Verlag, Berlin/Heidelberg/New York, pp. 181–191.

Dow-Edwards, D.L., Crane, A., Kennedy, C., and Sokoloff, L. (1982) Alterations of brain glucose metabolism in cretinism. *Soc. Neurosci. Abstr.* 8, 872.

Duffy, T.E., Cavazzuti, M., Cruz, N.F., and Sokoloff, L. (1982) Local cerebral glucose metabolism in newborn dogs: Effects of hypoxia and halothane anesthesia. *Ann. Neurol.* 11, 233–246.

Durham, D. and Woolsey, T.A. (1977) Barrels and columnar cortical organization: Evidence from 2-deoxyglucose (2-DG) experiments. *Brain Res.* 137, 169–174.

Eayrs, J.T. (1964) Endocrine influence on cerebral development. *Arch. Biol. Liege* 75, 529–565.

Eklöf, B., Lassen, N.A., Nilsson, L., Norberg, K., and Siesjö, B.K. (1973) Blood flow and metabolic rate for oxygen in the cerebral cortex of the rat. *Acta Physiol. Scand.* 88, 587–589.

Engel, J., Jr., Wolfson, L., and Brown, L. (1978) Anatomical correlates of electrical and behavioral events related to amygdaloid kindling. *Ann. Neurol.* 3, 538–544.

Fieschi, C., Lenzi, G. L., Zanette, E., Orzi, F., and Passarro, S. (1980) Effects on EEG of the osmotic opening of the blood-brain barrier in rats. *Life Sciences* 27, 239–243.

Foster, N.L., Chase, T.N., Fedio, P., Patronas, N.J., Brooks, R.A., and DiChiro, G. (1983) Alzheimer's disease: focal cortical changes shown by positron emission tomography. *Neurology* 33, 961–965.

Freygang, W.H. Jr. and Sokoloff, L. (1958) Quantitative measurement of regional circulation in the central nervous system by the use of radioactive inert gas. *Adv. Biol. Med. Phys.* 6, 263–279.

Friedli, C. (1978) Kinetics of changes in pO_2 and extracellular potassium activity in stimulated rat sympathetic ganglia. In I.A. Silver, M. Erecinska, and H.I. Bicher (eds.), *Adv. Exper. Med. Biol., Oxygen Transport to Tissue III*, Vol. 94. Plenum, New York, pp. 747–754.

Furlow, T. W., Hallenbeck, J. M., and Goodman, J. C. (1980) Adrenergic blocking agents modify the auditory-evoked response in the rat. *Brain Res.* 189, 269–273.

Galvan, M., Ten Bruggencate, G., and Senekowitsch, R. (1979) The effects of neuronal stimulation and ouabain upon extracellular K^+ and Ca^{2+} levels in rat isolated sympathetic ganglia. *Brain Res.* 160, 544–548.

Gjedde, A., Caronna, J.J., Hindfelt, B., and Plum, F. (1975) Whole-brain blood flow and oxygen metabolism in the rat during nitrous oxide anesthesia. *Amer. J. Physiol.* 229, 113–118.

Goochee, C., Rasband, W., and Sokoloff, L. (1980) Computerized densitometry and color coding of [^{14}C]deoxyglucose autoradiographs. *Ann. Neurol.* 7, 359–370.

Hand, P.J. (1981) The 2-deoxyglucose method. In L. Heimer and M.J. Robards (eds.), *Neuroanatomical Tract Tracing Methods.* Plenum, New York, pp. 511–538.

Hand, P.J., Greenberg, J.H., Miselis, R.R., Weller, W.L., and Reivich, M. (1978) A normal and altered cortical column: A quantitative and qualitative (^{14}C)-2 deoxyglucose (2 DG) mapping study. *Soc. Neurosci. Abstr.* 4, 553.

Hawkins, R., Phelps, M., Huang, S.C., and Kuhl, D. (1981) Effect of ischemia upon quantification of local cerebral metabolic rates for glucose with 2-(F-18)fluoro-deoxyglucose (FDG). *J. Cereb. Blood Flow Metab.* 1(Suppl 1), S9–S10.

Hers, H.G. (1957) *Le Métabolisme du Fructose.* Editions Arscia, Bruxelles, p. 102.

Horowicz, P. and Larrabee, M.G. (1958) Glucose consumption and lactate production in a mammalian sympathetic ganglion at rest and in activity. *J. Neurochem.* 2, 102–118.

Hosokawa, S., Iguchi, T., Caveness, W.F., Kato, M., O'Neill, R.R., Wakisaka, S., and Malamut, B.L. (1980) Effects of manipulation of the sensorimotor system on focal motor seizures in the monkey. *Ann. Neurol.* 7, 222–229.

Huang, S.C., Phelps, M.E., Hoffman, E.J., Sideris, K., Selin, C.J., and Kuhl, D.E. (1980) Noninvasive determination of local cerebral metabolic rate of glucose in man. *Am. J. Physiol.* 238, E69—E82.

Hubel, D.H. and Wiesel, T.N. (1968) Receptive fields and functional architecture of monkey striate cortex. *J. Physiol.* 195, 215–243.

Hubel, D.H. and Wiesel, T.N. (1972) Laminar and columnar distribution of geniculo-cortical fibers in the macaque monkey. *J. Comp. Neurol.* 146, 421–450.

Hubel, D.H., Wiesel, T.N., and Stryker, M.P. (1978) Anatomical demonstration of orientation columns in the macaque monkey. *J. Comp. Neurol.* 177, 361–380.

Kadekaro, M., Gross, P.M., Sokoloff, L., Holcomb, H.H., and Saavedra, J.M. (1983) Elevated glucose utilization in subfornical organ and pituitary neural lobe of the Brattleboro rat. *Brain Res.* 275, 189–193.

Kato, M., Malamut, B.L., Caveness, W.F., Hosokawa, S., Wakisaka, S., and O'Neill, R.R. (1980) Local cerebral glucose utilization in newborn and pubescent monkeys during focal motor seizures. *Ann. Neurol.* 7, 204–212.

Kennedy, C., Des Rosiers, M., Reivich, M., Sharp, F., Jehle, J.W., and Sokoloff, L. (1975) Mapping of functional neural pathways by autoradiographic survey of local metabolic rate with [^{14}C]deoxyglucose. *Science* 187, 850–853.

Kennedy, C., Des Rosiers, M.H., Sakurada, O., Shinohara, M., Reivich, M., Jehle, J.W., and Sokoloff, L. (1976) Metabolic mapping of the primary visual system of the monkey by means of the autoradiographic [^{14}C]deoxyglucose technique. *Proc. Natl. Acad. Sci. USA* 73, 4230–4234.

Kennedy, C., Sakurada, O., Shinohara, M., Jehle, J., and Sokoloff, L. (1978) Local cerebral glucose utilization in the normal conscious macaque monkey. *Ann. Neurol.* 4, 293–301.

Kennedy, C., Miyaoka, M., Suda, S., Macko, K., Jarvis, C., Mishkin, M., and Sokoloff, L. (1980) Local metabolic responses in brain accompanying motor activity. *Trans. Am. Neurol. Assoc.* 105, 13–17.

Kennedy, C., Gillin, J.C., Mendelson, W., Suda, S., Miyaoka, M., Ito, M., Nakamura, R.K., Storch, F.I., Pettigrew, K., Mishkin, M., and Sokoloff, L. (1982) Local cerebral glucose utilization in non-rapid eye movement sleep. *Nature* 297, 325–327.

Kety, S.S. (1950) Circulation and metabolism of the human brain in health and disease. *Am. J. Med.* 8, 205–217.

Kety, S.S. (1957) The general metabolism of the brain *in vivo*. In D. Richter (ed.), *Metabolism of the Nervous System*. Pergamon, London, pp. 221–237.

Kety, S.S. (1960) Measurement of local blood flow by the exchange of an inert, diffusible substance. *Methods Med. Res.* 8, 228–236.

Kety, S.S. and Schmidt, C.F. (1948a) The nitrous oxide method for the quantitative determination of cerebral blood flow in man: Theory, procedure, and normal values. *J. Clin. Invest.* 27, 476–483.

Kety, S.S. and Schmidt, C.F. (1948b) Effects of altered arterial tensions of carbon dioxide and oxygen on cerebral blood flow and cerebral oxygen consumption of normal young men. *J. Clin. Invest.* 27, 484–492.

Kuhl, D., Engel, J., Phelps, M., and Selin, C. (1979) Patterns of local cerebral metabolism and perfusion in partial epilepsy by emission computed tomography of ^{18}F-fluorodeoxyglucose and ^{13}N-ammonia. *Acta Neurol. Scand. [Suppl. 72]* 60, 538–539.

Kuhl, D.E., Engel, J. Jr., Phelps, M.E., and Selin, C. (1980) Epileptic patterns of local cerebral metabolism and perfusion in humans determined by emission computed tomography of ^{18}FDG and ^{13}NH$_3$. *Ann. Neurol.* 8, 348–360.

Kuhl, D.E., Phelps, M.E., Markham, C.H., Metter, E.J., Riege, W.H., and Winter, J. (1982) Cerebral metabolism and atrophy in Huntington's disease determined by ^{18}FDG and computed tomographic scan. *Ann. Neurol.* 12, 425–434.

Kuhl, D.E., Metter, E.J., Riege, W.H., Hawkins, R.A., Mazziotta, J.C., Phelps, M.E., and Kling, A.S. (1983) Local cerebral glucose utilization in elderly patients with depression, multiple infarct dementia, and Alzheimer's disease. *J. Cerebr. Blood Flow Metab.* 3(1), S494–S495.

Landau, W.M., Freygang, W.H. Jr., Rowland, L.P., Sokoloff, L., and Kety, S.S. (1955) The local circulation of the living brain; values in the unanesthetized and anesthetized cat. *Trans. Am. Neurol. Assn.* 80, 125–129.

Larrabee, M.G. (1958) Oxygen consumption of excised sympathetic ganglia at rest and in activity. *J. Neurochem.* 2, 81–101.

Lashley, K.S. (1934) The mechanism of vision. VII. The projection of the retina upon the primary optic centers of the rat. *J. Comp. Neurol.* 59, 341–373.

Lassen, N.A. (1959) Cerebral blood flow and oxygen consumption in man. *Physiol. Rev.* 39, 183–238.

Lassen, N.A. and Munck, O. (1955) The cerebral blood flow in man determined by the use of radioactive krypton. *Acta Physiol. Scand.* 33, 30–49.

Macko, K.A., Jarvis, C.D., Kennedy, C., Miyaoka, M., Shinohara, M., Sokoloff, L., and Mishkin, M. (1982) Mapping the primate visual system with [2-^{14}C]deoxyglucose. *Science* 218, 394–396.

Mata, M., Fink, D.J., Gainer, H., Smith, C.B., Davidsen, L., Savaki, H., Schwartz, W.J., and Sokoloff, L. (1980) Activity-dependent energy metabolism in rat posterior pituitary primarily reflects sodium pump activity. *J. Neurochem.* 34, 213–215.

McCulloch, J., Savaki, H.E., McCulloch, M.C., and Sokoloff, L. (1979) Specific distribution of metabolic alterations in cerebral cortex following apomorphine administration. *Nature* 282, 303–305.

McCulloch, J., Savaki, H.E., and Sokoloff, L. (1980a) Influence of dopaminergic systems on the lateral habenular nucleus of the rat. *Brain Res.* 194, 117–124.

McCulloch, J., Savaki, H.E., McCulloch, M.C., and Sokoloff, L. (1980b) Retina-dependent activation by apomorphine of metabolic activity in the superficial layer of the superior colliculus. *Science* 207, 313–315.

Miyaoka, M., Shinohara, M., Batipps, M., Pettigrew, K.D., Kennedy, C., and Sokoloff, L. (1979a) The relationship between the intensity of the stimulus and the metabolic response in the visual system of the rat. *Acta Neurol. Scand. [Suppl. 72]* 60, 16–17.

Miyaoka, M., Shinohara, M., Kennedy, C., and Sokoloff, L. (1979b) Alterations in local cerebral glucose utilization (LCGU) in rat brain during hypoxemia. *Trans. Am. Neurol. Assoc.* 104, 151–154.

Montero, V.M. and Guillery, R.W. (1968) Degeneration in the dorsal lateral geniculate nucleus of the rat following interruption of the retinal or cortical connections. *J. Comp. Neurol.* 134, 211–242.

Nelson, T. and Sokoloff, L. (1983) 2-Deoxyglucose incorporation into cerebral glycogen. *J. Neurochem.* 41(Suppl), S161A.

Nordmann, J.J. (1977) Ultrastructural morphometry of the rat neurohypophysis *J. Anat.* 123, 213–218.

Oldendorf, W.H. (1971) Brain uptake of radiolabeled amino acids, amines, and hexoses after arterial injection. *Amer. J. Physiol.* 221, 1629–1638.

Ornberg, R.L., Neale, E.A., Smith, C.B., Yarowsky, P., and Bowers, L.M. (1979) Radioautographic localization of glucose utilization by neurons in culture. *J. Cell Biol. Abstr.* 83, CN142A.

Pappius, H.M., Savaki, H.E., Fieschi, C., Rapoport, S.I., and Sokoloff, L. (1979) Osmotic opening of the blood–brain barrier and local cerebral glucose utilization. *Ann. Neurol.* 5, 211–219.

Phelps, M.E., Huang, S.C., Hoffman, E.J., Selin, C., Sokoloff, L., and Kuhl, D.E. (1979) Tomographic measurement of local cerebral glucose metabolic rate in humans with (F-18)2-fluoro-2-deoxy-D-glucose: Validation of method. *Ann. Neurol.* 6, 371–388.

Phelps, M.E., Kuhl, D.E., and Mazziotta, J.C. (1981) Metabolic mapping of the brain's response to visual stimulation: Studies in man. *Science* 211, 1445–1448.

Plum, F., Gjedde, A., and Samson, F.E. (eds.) (1976) Neuroanatomical functional mapping by the radioactive 2-deoxy-D-glucose method. *Neurosci. Res. Prog. Bull.* 14, 457–518.

Porrino, L., Namba, H., Crane, A., Jehle, J., and Sokoloff, L. (1982) Modulation of local cerebral glucose metabolism by estrogen and progesterone in the hypothalamus of ovariectomized female rats. *Soc. Neurosci. Abstr.* 8, 69.

Post, R.M., Kennedy, C., Shinohara, M., Squillace, K., Miyaoka, M., Suda, S., Ingvar, D.H., and Sokoloff, L. (1979) Local cerebral glucose utilization in lidocaine-kindled seizures. *Soc. Neurosci. Abstr.* 5, 196.

Pulsinelli, W.A. and Duffy, T.E. (1979) Local cerebral glucose metabolism during controlled hypoxemia in rats. *Science* 204, 626–629.

Rakic, P. (1976) Prenatal genesis of connections subserving ocular dominance in the rhesus monkey. *Nature* 261, 467–471.

Reivich, M., Jehle, J., Sokoloff, L., and Kety, S.S. (1969) Measurement of regional cerebral blood flow with antipyrine-^{14}C in awake cats. *J. Appl. Physiol.* 27, 296–300.

Reivich, M., Kuhl, D., Wolf, A., Greenberg, J., Phelps, M., Ido, T., Cassella, V., Fowler, J., Hoffman, E., Alavi, A., Som, P., and Sokoloff, L. (1979) The [^{18}F]fluoro-deoxyglucose method for the measurement of local cerebral glucose utilization in man. *Circ. Res.* 44, 127–137.

Roth, R.H. (1976) Striatal dopamine and gamma-hydroxybutyrate. *Pharmacol. Ther.* 2, 71–88.

Roth, R.H. and Giarman, N.J. (1966) γ-Butyrolactone and γ-hydroxybutyric acid—I. Distribution and metabolism. *Biochem. Pharmacol.* 15, 1333–1348.

Sacks, W. (1957) Cerebral metabolism of isotopic glucose in normal human subjects. *J. Appl. Physiol.* 10, 37–44.

Sakurada, O., Shinohara, M., Klee, W.A., Kennedy, C., and Sokoloff, L. (1976) Local cerebral glucose utilization following acute chronic morphine administration and withdrawal. *Soc. Neurosci. Abstr.* 2, 613.

Savaki, H.E., Kadekaro, M., Jehle, J., and Sokoloff, L. (1978) α- and β-adrenoreceptor blockers have opposite effects on energy metabolism of the central auditory system. *Nature* 276, 521–523.

Savaki, H.E., McCulloch, J., Kadekaro, M., and Sokoloff, L. (1982) Influence of α-receptor blocking agents upon metabolic activity in nuclei involved in central control of blood pressure. *Brain Res.* 233, 347–358.

Scheinberg, P. and Stead, E.A. Jr. (1949) The cerebral blood flow in male subjects as measured by the nitrous oxide technique. Normal values for blood flow, oxygen utilization, and peripheral resistance, with observations on the effect of tilting and anxiety. *J. Clin. Invest.* 28, 1163–1171.

Schuier, F., Orzi, F., Suda, S., Kennedy, C., and Sokoloff, L. (1981) The lumped constant for the [^{14}C]deoxyglucose method in hyperglycemic rats. *J. Cereb. Blood Flow Metab.* 1(1), S63.

Schwartz, W.J. (1978) A role for the dopaminergic nigrostriatal bundle in the pathogenesis of altered brain glucose consumption after lateral hypothalamic lesions. Evidence using the ^{14}C-labeled deoxyglucose technique. *Brain Res.* 158, 129–147.

Schwartz, W.J., Sharp, F.R., Gunn, R.H., and Evarts, E.V. (1976) Lesions of ascending dopaminergic pathways decrease forebrain glucose uptake. *Nature (Lond.)* 261, 155–157.

Schwartz, W.J. and Gainer, H. (1977) Suprachiasmatic nucleus: Use of ^{14}C-labeled deoxyglucose uptake as a functional marker. *Science* 197, 1089–1091.

Schwartz, W.J., Smith, C.B., Davidsen, L., Savaki, H., Sokoloff, L., Mata, M., Fink, D.J., and Gainer, H. (1979) Metabolic mapping of functional activity in the hypothalamo-neurohypophysial system of the rat. *Science* 205, 723–725.

Schwartz, W.J., Davidsen, L.C., and Smith, C.B. (1980) *In vivo* metabolic activity of a putative circadian oscillator, the rat suprachiasmatic nucleus. *J. Comp. Neurol.* 189, 157–167.

Sejnowski, T.J., Reingold, S.C., Kelley, D.B., and Gelperin, A. (1980) Localization of [^{3}H]-2-deoxyglucose in single molluscan neurones. *Nature* 287, 449–451.

Sharp, F.R., Kauer, J.S., and Shepherd, G.M. (1975) Local sites of activity-related glucose metabolism in rat olfactory bulb during olfactory stimulation. *Brain Res.* 98, 596–600.

Shinohara, M., Sakurada, O., Jehle, J., and Sokoloff, L. (1976) Effects of D-lysergic acid diethylamide on local cerebral glucose utilization in the rat. *Soc. Neurosci. Abstr.* 2, 615.

Shinohara, M., Dollinger, B., Brown, G., Rapoport, S., and Sokoloff, L. (1979) Cerebral glucose utilization: Local changes during and after recovery from spreading cortical depression. *Science* 203, 188–190.

Silverman, M.S., Hendrickson, A.E., and Clopton, B.M. (1977) Mapping of the tonotopic organization of the auditory system by uptake of radioactive metabolites. *Soc. Neurosci. Abstr.* 3, 11.

Smith, C.B., Goochee, C., Rapoport, S.I., and Sokoloff, L. (1980) Effects of ageing on local rates of cerebral glucose utilization in the rat. *Brain* 103, 351–365.

Smith, C.B. (1983) Localization of activity-associated changes in metabolism of the central nervous system with the deoxyglucose method: Prospects for cellular resolution. In J.L. Barker and J.F. McKelvy (eds.), *Current Methods in Cellular Neurobiology.* John Wiley & Sons, New York, pp. 269–317.

Sokoloff, L. (1960) Metabolism of the central nervous system *in vivo.* In J. Field, H.W. Magoun, and V.E. Hall (eds.), *Handbook of Physiology-Neurophysiology,* Vol III. American Physiological Society, Washington, D.C., pp. 1843–1864.

Sokoloff, L. (1966) Cerebral circulatory and metabolic changes associated with aging. *Res. Publ. Assoc. Nerv. Ment. Dis.* 41, 237–254.

Sokoloff, L. (1969) Cerebral circulation and behavior in man: Strategy and findings. In A.J. Mandell and M.P. Mandell (eds.), *Methods and Theory in Psychochemical Research in Man.* Academic Press, New York, pp. 237–252.

Sokoloff, L. (1976) Circulation and energy metabolism of the brain. In G.J. Siegel, R.W. Albers, R. Katzman, and B.W. Agranoff (eds.), *Basic Neurochemistry,* Second Edition. Little, Brown & Co., Boston, pp. 388–413.

Sokoloff, L. (1977) Relation between physiological function and energy metabolism in the central nervous system. *J. Neurochem.* 29, 13–26.

Sokoloff, L. (1978a) Mapping cerebral functional activity with radioactive deoxyglucose. *Trends Neurosci.* 1(3), 75–79.

Sokoloff, L. (1978b) Local cerebral energy metabolism: Its relationships to local functional activity and blood flow. In M.J. Purves and K. Elliott (eds.), *Cerebral Vascular Smooth Muscle and Its Control,* Ciba Foundation Symposium 56. Elsevier/Excerpta Medica/North-Holland, Amsterdam, pp. 171–197.

Sokoloff, L. (1979) The [^{14}C]deoxyglucose method: Four years later. *Acta Neurol. Scand. [Suppl. 70]* 60, 640–649.

Sokoloff, L. (1982) The radioactive deoxyglucose method: Theory, procedure, and applications for the measurement of local glucose utilization in the central nervous system. In B.W. Agranoff and M.H. Aprison (eds.), *Advances in Neurochemistry,* Vol. 4. Plenum, New York, pp. 1–82.

Sokoloff, L., Wechsler, R.L., Mangold, R., Balls, K., and Kety, S.S. (1953) Cerebral blood flow and oxygen consumption in hyperthyroidism before and after treatment. *J. Clin. Invest.* 32, 202–208.

Sokoloff, L., Reivich, M., Kennedy, C., Des Rosiers, M.H., Patlak, C.S., Pettigrew, K.D., Sakurada, O., and Shinohara, M. (1977) The [^{14}C]deoxyglucose method for the measurement of local cerebral glucose utilization: Theory, procedure, and normal values in the conscious and anesthetized albino rat. *J. Neurochem.* 28, 897–916.

Sols, A. and Crane, R.K. (1954) Substrate specificity of brain hexokinase. *J. Biol. Chem.* 210, 581–595.

Suda, S., Shinohara, M., Miyaoka, M., Kennedy, C., and Sokoloff, L. (1981) Local cerebral glucose utilization in hypoglycemia. *J. Cerebr. Blood Flow Metab.* 1(1), S62.

Webster, W.R., Serviere, J., Batini, C., and LaPlante, S. (1978) Autoradiographic demonstration with 2-[^{14}C]deoxyglucose of frequency selectivity in the auditory system of cats under conditions of functional activity. *Neurosci. Lett.* 10, 43–48.

Wechsler, L.R., Savaki, H.E., and Sokoloff, L. (1979) Effects of *d*- and *l*-amphetamine on local cerebral glucose utilization in the conscious rat. *J. Neurochem.* 32, 15–22.

Wiesel, T.N., Hubel, D.H., and Lam, D.M.K. (1974) Autoradiographic demonstration of ocular dominance columns in the monkey striate cortex by means of transneuronal transport. *Brain Res.* 79, 273–279.

Wolfson, L.I., Sakurada, O., and Sokoloff, L. (1977) Effects of γ-butyrolactone on local cerebral glucose utilization in the rat. *J. Neurochem.* 29, 777–783.

Yarowsky, P.J., Crane, A.M., and Sokoloff, L. (1980) Stimulation of neuronal glucose utilization by antidromic electrical stimulation in the superior cervical ganglion of the rat. *Soc. Neurosci. Abstr.* 6, 340.

Yarowsky, P., Kadekaro, M., and Sokoloff, L. (1983) Frequency-dependent activation of glucose utilization in the superior cervical ganglion by electrical stimulation of cervical sympathetic trunk. *Proc. Natl. Acad. Sci. USA* 80, 4179–4183.

INDEX

ATPase, 50, 66
Auditory cortex
  cerebral glucose utilization
    rate, 25, 28, 32
    effect of amphetamine, 60
    under anesthesia, 29
  rate constant value, 10–11
Auditory occlusion, 32
Auditory system
  cerebral glucose utilization
    in, 25, 28, 32
    aging and, 71
    effect of adrenergic
      blocking agents, 64
  thyroid function, 55
  tonotopic representation
    studies, 46
Autoradiographs
  color coding of, 27–28, 48,
    color plate
  dopaminergic pathways,
    62–63
  experimental techniques, 18
  focal seizures, 33
  neurohypophysial studies, 52
  spreading cortical depression,
    67–69
  visual occlusion, 35–39

Barbiturate anesthesia, 31,
  67–69
Basal ganglia
  cerebral glucose utilization
    rate, 25
    during focal seizure, 65
Beagle, *see* Dogs
Bilateral symmetry, of visual
  system, 35, 36–37
Blind spots, 38, 40, 42
Blood–brain barrier, 5, 20
  opening of, 66
Brain sections, experimental

  preparation, 17–18
Brain stem, 25
Brattleboro rat, 54
γ-Butyrolactone, 56–57

Cats
  auditory system studies, 46
  cerebral glucose utilization
    rate, 25–26
  experimental procedures, 18
  lumped constant value, 14
  rate constant values, 9
Catheters, 16, 17
Caudate
  cerebral glucose utilization
    rate, 25
    aging and, 71
    during focal seizure, 32
    effect of amphetamine, 60
Caudate–putamen
  cerebral glucose utilization
    rate
    effect of lesions, 58
    under anesthesia, 30
  rate constant value, 10–11
Cellular compartments, 25–26
Cerebellar cortex
  cerebral glucose utilization
    rate, 26
    during focal seizure, 65
    under anesthesia, 30
  rate constant value, 10–11
Cerebellar flocculus, 61
Cerebellar hemispheres, 61
Cerebellar nuclei
  cerebral glucose utilization
    rate
    effect of amphetamine, 61
    under anesthesia, 30
  rate constant value, 10–11
Cerebellar white matter
  cerebral glucose utilization

"Mirror" focus, 32
Monkeys
    circadian rhythm studies, 47
    determination of lumped
        constant, 9, 12–13
    experimental procedures, 18
    focal seizure studies, 64–65
    motor function studies, 46
    rate constant values, 9
    sleep studies, 47
    see also Rhesus monkey
Morphine, 57–58
Motor cortex, 32, 33, 65
Motor functions, 46

Naloxone, 58
Nauta degeneration methods, 34
Neuroendocrinology, 51–56
Neurohypophysial system,
    51–54
Neuropathophysiology, 64–72
Neuropharmacology, 56–64
Neuropil, 71
Nissl staining, 52–53
Nitrogen, 70
Nitrous oxide, 17, 70
Nitrous oxide technique, 2
Norway brown rats, visual
        system studies, 42–45
Nucleus accumbens, 61

Occipital cortex, human, 74–75
Ocular dominance columns,
        34–35, 37, 38, 42
Oculomotor nuclei, 36–37
Olfactory bulb, 46
Olfactory cortex, 60
Olfactory nuclei, 60
Olfactory system, studies of, 46
Operational equation, 7, 8, 15,
    20–22

constants and errors, 19
Optic chiasma, 34

Paraventricular nucleus, 52–54
Parietal cortex
    cerebral glucose utilization
        rate, 25
    effect of amphetamine, 60
    spreading cortical
        depression, 65–69
    under anesthesia, 29
    rate constant value, 10–11
Pars compacta, 58
Pasteur effect, 70
Pathophysiology, 64–72
Penicillin-induced focal
        seizures, 31–32, 33,
        64–65
Phenoxybenzamine, 52–54
Phentolamine, 64
Pigmentation, visual system
        studies, 43–45
Pituitary, 52–54
Pontine gray
    cerebral glucose utilization
        rate
        effect of amphetamine, 61
        under anesthesia, 30
    rate constant value, 10–11
Pontine reticular nucleus, 26
Positron emission tomography,
        26, 73
Posterolateral thalamic nucleus,
        44
Potassium ions, 50
    spreading cortical depression
        and, 65–69
Precursor pool, 8, 21
    half-life of, 11
    operational equation and, 9,
        13–14

Thiopental, 29–30
Thyroid, 54–55
Toluidine blue staining, 52–53
Tranquilization, 3

UDP–deoxyglucose, 5

Vasopressin, 53–54
Vestibular nucleus
  cerebral glucose utilization
    rate, 25
    effect of amphetamine, 61
    under anesthesia, 30
Visual cortex
  cerebral glucose utilization
    rate
    effect of amphetamine, 61
    under anesthesia, 29
    visual occlusion and,
      34–40
  rate constant value, 10–11
  visual system studies, 43, 45
Visual occlusion, 34–40

Visual system, 34–39, 42–45
  autoradiographs, 62–63
  human, 74–75
  dopaminergic pathways,
    58–59, 62–63
  pigmentation and, 43–45

Weber-Fechner Law, 43
"Whisker barrels," 46, *color
  plate*
White matter
  cerebral glucose utilization
    rates, 26, 28, 29–30
    effect of amphetamine, 61
    effect of γ-butyrolactone,
     57
    effect of hypoxemia, 70–71
    effect of morphine, 58
    rate constant value, 10–11
Withdrawal syndrome, 58

X-ray film, 18

The Magnes Lecture Series is designed
by Martin Lubin of Betty Binns Graphics.

This book was set in Bodoni Book
at David E. Seham Associates, Inc.
Joseph J. Vesely coordinated production.